Minerals, Fossils, and Fluorescents of Arizona

A Field Guide for Collectors

Neil R. Bearce

ARIZONA DESERT ICE PRESS
Tempe, Arizona

©2003, 2006 by Neil R. Bearce
with the exception of National Geographic Topo! maps
and the Arizona Geological Society Arizona Rocks
and Physiographic maps.

Published by
Arizona Desert Ice Press Inc.
P.O. Box 27347
Tempe, AZ 85285

All rights reserved. No part of this book may be reproduced by any mechanical, photographic, or electronic process, or in the form of a phonographic recording, nor may it be stored in a retrieval system, transmitted, or otherwise copied for public or private use, without written permission of the publisher.

Printed in the United States of America
10 9 8 7 6 5 4 3 2 1

Book design and production by Harrison Shaffer

Color plates photographed by Jeffery Scovil

Library of Congress Control Number 2005936752

ISBN-13: 978-0-9749846-1-2
ISBN-10: 0-9749846-1-2

The author, publisher, and all parties involved in the writing, production, and sale of this book disclaim any responsibility for the safety and legality of the reader's activities. The routes of travel and the collecting sites mentioned in this book are subject to change without notice. Knowing the condition of these routes and sites and determining if they can be safely and legally traveled and visited is the sole responsibility of the reader. The mention of a location in this book does not constitute permission to enter or collect.

☙
To Charli
with love
☙

Contents

	List of Tables and Illustrations	xi
	Preface and Acknowledgments	xiii
	Introduction	1
	Rockhound Etiquette	5
	Scientific Properties of Site Specimens	7
	Minerals	27
	THE MINERAL COLLECTING SITES	
1.	Selenite near Dutchman Wash	47
4.	Hematite at the BBC Mine	50
5.	Cummingtonite East of the BBC Mine	53
6.	Chrysocolla along the Rankin-Lincoln Ranch Road	55
7.	Malachite at the Gold Hill Mine	58
8.	Augite at the Green Streak Mine	60
9.	Minerals South of Plomosa Road	63
10.	Barite West of Plomosa Road	67
11.	Epidote near Boyer Gap	69
12.	Alunite on Sugarloaf Peak	73
13.	Bladed Hematite at the Big Bertha Extension Mine	76
14.	Geodes East of the Ramsey Mine	78
15.	Chalcedony on Hull Road	81
16.	Malachite at the Mammoth Mine	84
17.	Hematite and Copper Minerals at Cunningham Pass	87
18.	Copper Minerals at the Bullard Mine	90
19.	Amethyst at the Contact Mine	93
20.	Fluorite and Selenite off Eagle Eye Road	96
24.	Jasper and Hematite at Mingus Mountain	101

25.	Siderite at the Copper Chief Mine	104
26.	Travertine at the Empire Onyx Quarries	107
27.	Glauberite Pseudomorphs along Salt Mine Road	111
29.	Calcite Near Forest Road 41	115
30.	Banded Jasper off Table Mesa Road	118
32.	Mica at Cottonwood Gulch	121
33.	Agate at Signal City	124
34.	Banded Agate at the McCraken Mine	128
35.	Chalcedony Lined Geodes near Keenan Camp	132
36.	Fluorite and Selenite at the Lead Pill Mine	135
37.	Malachite at the New England Mine	138
39.	Copper Minerals at the Cactus Queen Mine	141
46.	Obsidian on the Barry M. Goldwater Range – Area A	144
47.	Chalcedony and Geodes on the Barry M. Goldwater Range – Area B	148
49.	Chert at Battleground Ridge	153
52.	Pegmatites along the Black Pearl Road	156
53.	Malachite at the United States Mine	159
57.	Marble off Ryan Ranch Road	162
58.	Geodes at Corral Nuevo	165
60.	Chrysotile at the Phillips Mine	168
61.	Onyx on Forest Road 303	173
62.	Serpentine on Forest Road 189	176
64.	Amethyst at the Woodpecker Mine	180
68.	Vanadinite at the Grey Horse Mine	183
70.	The Copper Creek District	186
72.	Selenite Roses East of Saint David	192
74.	Marble on Forest Road 689	195
75.	Minerals in the Tombstone Hills	198
77.	Quartz Crystal Druse at Gold Gulch	202
79.	Porphyry at the W.A. Ranch Well	205
81.	Chalcedony On Route to Coyote Spring	208
82.	Geodes South of Ash Peak	211
83.	Opal on State Route 75	215
84.	Minerals at the Carlisle dumps	218

86.	Dendrites Near Steins, New Mexico	221
	Fossils	253
	THE FOSSIL COLLECTING SITES	
2.	Mollusca in Nail Canyon	269
3.	Crinoidea in Limestone Canyon	272
21.	Plant Fossils at D K Well	275
22.	Marine Fossils west of Jerome	278
23.	Crinoidea off State Route 89A	281
28.	Colonial Coral near Chasm Creek	284
48.	Mollusca at Battleground Ridge	287
50.	Crinoidea on Forest Road 29A	290
51.	Porifera, Braciopoda, and Crinoidea along Forest Road 237	293
54.	Colonial Coral South of Control Road	296
55.	Coral Fossils South of Forest Road 144	299
56.	Crinoidea off Forest Road 29	301
59.	Rugosa off Forest Road 138	304
67.	Trace Fossils at Dago Spring	308
69.	Bivalva on State Route 77	310
73.	Marine Fossils in French Joe Canyon	313
76.	Gastropoda in the Mule Mountains	316
89.	Gastropoda at the Cochise Mine	319
90.	Crinoidea at the Willie Rose Mine	322
	Fluorescents	325
	THE FLUORESCENT COLLECTING SITES	
31.	Fluorescent Calcite near Cottonwood Gulch	333
38.	Fluorescent Geodes and Chalcedony at the Maggie Mine	336
40.	Fluorescent Calcite at the Scott and Black Pearl Mines	340
41.	Fluorescent Calcite at the Black Silver Mine	344
42.	Fluorescents West of Morristown	347
43.	Fluorescent Calcite Near Twin Buttes	355
44.	Fluorescent Calcite at the Prince Mine	358
45.	Fluorescent Calcite North of Lime Hill	361
63.	Fluorescents South of Cottonwood Canyon Road	365
65.	Fluorescents at the Ajax Mine	368

66.	Fluorescent Minerals South of Mineral Mountain	370
71.	Fluorescent Scheelite at the Tungsten King Mine	375
78.	Fluorescent Calcite at the Crook Tunnel	379
80.	Fluorescents Along the San Francisco River Road	382
85.	Fluorescent Calcite at the Summit Mine	385
87.	Fluorescent minerals off U.S. Route 80	387
88.	Fluorescent Minerals at the Hilltop Mine	390
	Index	393
	About the Author	399

Tables and Illustrations

TABLES

	Site difficulty Scale	24
1.	Color Streak	28
2.	Luster	29
3.	The Crystal Systems	30
4.	Crystal Habits	31
5.	The Mohs Hardness Scale and Rosiwal Cutting Scale	33
6.	Fracture Patterns	33
7.	Tenacity	34
8.	Density Comparison of Some Well Known Minerals	34
9.	Igneous Rock Mineralogy	37
10.	Sedimentary Rock Mineralogy	38
11.	Metamorphic Rock Protoliths	40
12.	Metamorphic Minerals of Interest to Collectors	40
13.	The Dana System of Mineral Classification	41
14.	Selected Arizona Sedimentary Formations	258
15.	Chronology of Life on Earth	260
16.	Fossil Environments	265
17.	Common Fluorescent Minerals	329

MAPS

1.	Physiographic Map of Arizona	4
2.	Arizona Collecting Site Location Map	25
3.	Arizona Rocks Map and Key	227

COLOR PLATES

1.	Fluorescents I	228
2.	Fluorescents II (Site 42 – West of Morristown)	230
3.	Fluorescents III	232
4.	Fluorescents IV	234
5.	Fluorescents V	236
6.	Fossils I	238

7.	Fossils II (Site 21 – Plant Fossils at DK Well)	239
8.	Fossils III	240
9.	Fossils IV	241
10.	Fossils V	242
11.	Fossils VI	243
12.	Minerals I	244
13.	Minerals II (Site 9 - South of Plomosa Road)	245
14.	Minerals III	246
15.	Minerals IV	247
16.	Minerals V	248
17.	Minerals VI (Along Signal and Alamo Roads)	249
18.	Minerals VII	250
19.	Minerals VIII	251
20.	Minerals IX	252

Preface

THIS IS A BOOK FOR AMATEURS (Latin for lovers). It is written for amateur mineralogists, paleontologists, and rockhounds who are inspired by the treasures of the natural world. Since prehistoric times, we have been spiritually as well as physically united with our natural surroundings. The natural world is our home and we love it. And why not? All of our life-sustaining needs are provided by it. But, we are at the same time fascinated, mystified, and perplexed by it. Unlike all our fellow creatures that inhabit the Earth with us, only we *Homo sapiens* (Latin for "wise man") cannot find contentment simply living in the world. No, we itch to understand it, and enjoy it, and organize it, and speculate about it. Curiosity may kill the cat, but for mankind it provides the stimulus for the aesthetic and scientific inquiry that differentiates us from all other living things on the planet.

Early on, our ancestors realized that our environment was composed of three zones; the atmosphere, the hydrosphere, and the lithosphere (Greek for air, water, and rock globes). Within these domains, our forebears categorized everything they encountered into three categories; the plant, animal, and mineral kingdoms. We reign supreme over all of these kingdoms using them for food, clothing, and shelter. If we cannot hunt it, we grow it. If we cannot grow it, we mine it. We have even learned to "improve" upon nature by domesticating and hybridizing some plants and animals and synthesizing certain minerals.

Our forefathers were not only hunter-gatherer wanderers, they were also wonderers. Once the hunting and gathering was done and they were adequately nourished, clothed, and sheltered, they took time to collect, possess, and contemplate the most interesting and aesthetically pleasing rocks, minerals, and fossils that they encountered along the way. We know this because we find these materials in their ancient dwellings and graves. They fashioned minerals and fossils into statuary and jewelry for adornment and ceremonial use. Continuing in that ancient tradition, once the hybridizing and synthesizing is done, we too can take some time off to collect, study, and enjoy the minerals and fossils that our Arizona environment has to offer. Happy hunting!

Acknowledgments

Thanks to the many rockhounds, cowboys, miners, bartenders, and others who I met at various rock shows, in little Arizona towns, in the field, and along the way for sharing their knowledge of collecting sites with me. George Stevens, Walter Johnson, Stan and Sue Celestian, Ray Grant, Pat McMahan, Jerry Blain among others provided information that resulted in the inclusion of collecting sites in this book. Nyal Neimuth's assistance with geological, mining history, and mineral identification questions is much appreciated. Thanks to Jeff Scovil for donating the cover photograph and to Ed Davis for contributing the excellent Hogan Mine fluorescent photograph that appears on the back cover. Love to Charli for the hardware upon which this book was written and for her patient assistance in showing me how to use it. Special thanks to National Geographic Maps for its kind permission to use its Topo! 2002 software from which the site maps in this book were made. Special thanks also to the Arizona Geological Survey for its kind permission to use the Arizona Rocks and Physiographic maps in this book.

Neil R. Bearce

Minerals, Fossils, and Fluorescents of Arizona

Introduction

IN 1997, I COMMITTED to write a book that became known as *Minerals of Arizona: A Field Guide for Collectors*. That undertaking was motivated by the Arizona rockhound community that was clamoring for a new, up to date field guide. I was a novice author and, in the finest novice tradition, leapt in where angels feared to tread. After expending huge amounts of time, money, and gasoline, I finally went to press in 1999 having nothing but faith and hope that enough people would buy the book to avert financial disaster. And indeed you did. Thank you very much!

You bought. You went. You found. But rockhounds are never satisfied. Readers reported to me that they had gone to one location or another described in the book, had an enjoyable outing, found good stuff, and asked when I was going to write another book. The clamoring never ceases because, despite our scientific advancement, we are still hunter-gatherers at heart. Mineral and fossil hunting not only gives us a reason to get out into remote desert and mountain retreats where we can get plenty of fresh air and exercise, but it also provides us the opportunity to reconnect with our prehistoric origins. Fossils and minerals can easily be purchased, but they never have as much meaning as the ones that Mother Nature gifts to us. The pleasure we derive from unearthing a treasure from her storehouse ourselves adds value far exceeding even the most expensive gem we can buy in the marketplace.

The theme of this book is the same as *Minerals of Arizona:* places you can go and rocks you can find. However, as indicated in the title, the scope of the subject matter is expanded to include fossils and fluorescents in addition to minerals. There are several reasons for this. First and most important, readers and critics expressed a need for such an expansion. Second, experience shows that people who are addicted to minerals are only a short step away from getting hooked on fossils and fluorescents as well. Rockhounds thirst for new locations that will yield unusual materials and afford new experiences. Finally, and unfortunately, mineral locations are becoming scarcer either because of changing land use requirements or because the supply of collectable material has been exhausted. Fossils and fluorescents open up new collecting opportunities because they are geographically more widely disbursed and are, generally, in greater supply. Even though *Minerals of Arizona* is only about five years old, it is becoming somewhat dated because of Arizona's changing landscape. For example, the onyx at Mayer (site 2) on State Route 69 is now inaccessible because the new road has completely covered up the once exposed vein. Agate on the Kofa-Manganese Road (site 20) is no longer available. The Kofa National Wildlife Refuge authority has decided that picking up rocks is inconsistent with its raison d'etre. Likewise, the green quartz south of State Route 238 (site 26) is now "protected" in the new Sonoran Desert National Monument by Presidential decree without the hindrance of public notice, hearing, or debate. The Blue Ball Mine (site 33) material is all but exhausted. And, the Hilton Mine (site 46) is under private ownership and the owner requests no trespassing.

Changes such as those listed above are inevitable and continual. Old quarries are filled in, old mines are made environmentally correct, and old collecting grounds are subdivided. It is regrettable to see a favorite old site and its unique material disappear. However, the most exciting rockhound experience is not at a well known picked over site, but at a new, heretofore undiscovered location. Fortunately, Arizona covers 113,956 square miles of deserts, mountains, canyons, basins, and grasslands. Since the Proterozoic Eon, over 570 million years ago, Arizona's landscape has been sculpted by climatic and geologic forces into many beautifully varied, scenic, and mineral and fossil rich geographic landforms. Geologically, the state is divided into three physiographic provinces that run diagonally across the state from the Northwest to the Southeast. The Colorado Plateau is the northernmost province, the Central Mountain, or Transition, Province is in the middle, and the Basin and Range is the southernmost province (see map on page 4). All of these provinces have been formed by a succession of tectonic uplift and subsidence of the earth's crust, erosion, physical and chemical weathering, volcanism, and the inundation and retreat of at least five ancient seas. Not all of Arizona's territory is open to prospecting but enough of it is to keep the average rockhound happy for many years to come. Your prospecting efforts may or may not result in the discovery of pretty rocks and interesting fossils. But in any case, be sure to enjoy the beauty and grandeur of Arizona's diverse landscapes.

The Colorado Plateau Province is the northern third of Arizona. It extends southward from the Utah border to the Mogollon Rim. The rock type here is primarily sedimentary punctuated by Cenozoic volcanic intrusions such as the San Francisco and White Mountains. Erosion has carved remarkable formations such as the Grand Canyon, the Painted Desert, and Monument Valley. Mineral collecting in this province is sparse compared to the other two. Much of the area is national park and Indian reservation both of which are off limits to collecting. However, petrified wood and marine fossils occur over a wide area between State Route 260 and Interstate 40. The Arizona Strip between the north rim of the Grand Canyon and the Utah and Nevada borders contains fossils and selenite.

The Transition Province is an area of rugged, jumbled, and contorted Precambrian and Paleozoic rock of various kinds. This is Arizona's copper belt. Mineralization here is very diverse as evidenced by the area's extensive mining history. Fossil hunting is also productive here especially in the vicinity of the Mogollon Rim where the Colorado Plateau has been severely eroded in the Tonto and Coconino National Forests and in the Verde Valley.

The Basin and Range Province is in the southern part of the state. It is characterized by a series of elongated mountain ranges oriented in a more or less northwest to southeast direction across the Sonoran Desert. The basins between the ranges are filled with alluvial deposits eroded from these ranges. Rock here is primarily Precambrian and Cenozoic. The mountains in this province are generally older tilted blocks or younger volcanoes. Because the rock here is more igneous and metamorphic than on the Colorado Plateau, mineralization is greater, but fossilization is much less.

Whatever province you are in, enjoy Arizona and find pretty rocks, fascinating fossils, and spectacular fluorescents.

Physiographic Map of Arizona

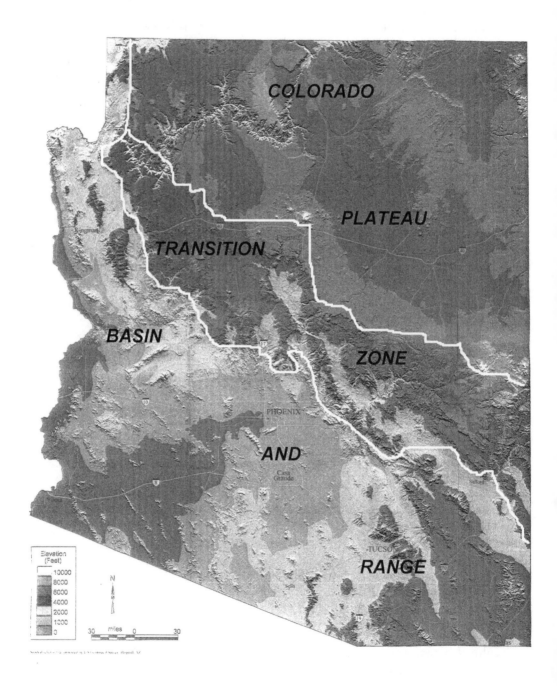

Rockhound Etiquette

THE FUTURE OF ROCKHOUNDING depends upon access to the land both public and private. Access to the land depends upon the reputation rockhounds cultivate for themselves. Public land managers, private landowners, ranchers, miners, and others who are in a position to influence land use and accessibility have little regard for people who make nuisances of themselves. Therefore, it benefits all of us who enjoy Arizona's great outdoors to be considerate and act responsibly. By adhering to the following practices, you will do your part to assure that rockhounds continue to be welcomed by those who regulate the land and by those whose livelihood depends upon it.

1. Respect private property and mining claim holder's rights. Ask permission to enter. People are often so surprised by your courtesy that they say yes!

2. When on public lands, obey the rules set by state and federal jurisdictions. Pay particular attention to posted notices concerning closed areas, wilderness areas, wildlife management areas, tribal lands, state trust land, etc.

3. Be considerate of others who also enjoy using the land such as hunters, hikers, campers, birdwatchers, etc.

4. Leave gates as you found them, usually closed, but not always. Stay clear of watering holes so that cattle will come in to drink.

5. Unhappily, much of our landscape is polluted and littered with trash. This is especially true of mining districts. Let's not add to it.

6. Vandalism of buildings, road signs, fences, and equipment is also a blight to be avoided.

7. Desert environments are fragile and the destructive effects of off-road joyriding take years to heal. Stay on existing roads.

8. It is worth saying again: ***"Only you can prevent forest fires."*** This applies to desert fires as well. Be sure your campfire is dead out.

9. Do not disturb archaeological sites or take artifacts. It is unlawful. A volunteer organization of over 100 site stewards quietly monitors historic landmarks statewide. They have provided information to law enforcement

authorities leading to the arrest and conviction of many who chose to scoff at antiquities laws.

10. Hammers, chisels, prybars, and other hand tools are common and acceptable tools for the rockhound hobbyist. Explosives and power equipment, however, are the instruments of professional miners.

11. Remember those who will come after you. Collect only the amount you can reasonably use and leave the rest for your fellow rockhounds to enjoy in the future.

12. If you dig in the ground, be sure to fill in your holes to prevent erosion.

13. Be legal. According to various U.S. Government publications and pamphlets, we may not collect in national parks and monuments or on prohibited public lands. We may collect for personal use but not for resale or barter without a permit. You may apply for a Mineral Material Sale Permit at your local BLM field office. We may collect "reasonable" amounts of minerals, rocks, and invertebrate fossils. Reasonable is defined as no more than 25 pounds of material per day plus one piece not to exceed 250 pounds per year. One person can not transfer his allotment to another. We may not collect using motorized, mechanical, or explosive devices. We may not unnecessarily degrade public lands while removing specimens. We may not collect vertebrate fossils without a permit. Unless you are a scientist or educator willing to donate your T-Rex to a museum, do not even think about applying for one. This is by no means a comprehensive list of all the rules and regulations that apply to all public lands. For more information, see The pamphlet entitled *Rockhounding in Arizona* published by the U.S. Department of the Interior, Bureau of Land Management. Check with your local land manager for rules specific to the area in which you plan to collect.

Scientific Properties of Site Specimens

AGATE ag'ət Greek *achatēs* (see quartz).

ALUNITE "aluminilite" 'al-(y)ə –nīt (Potassium aluminum sulfate hydroxide) Latin *aūlmen* for alumstone.

Environment:	A hydrothermal formation in the rocks of sulfide ore bodies.
Chemical Formula:	$KAl_3(SO_4)_2(OH)_6$.
Hardness:	3.5–4.0.
Color:	Colorless, white, gray, yellowish.
Streak:	White.
Crystal System and Habit:	Trigonal. Rhombohedral, columnar, tabular, massive.
Cleavage and Fracture:	Good.
Optical Properties:	Pearly. Opaque.
Specific Gravity:	2.7–2.8.
Tenacity:	Brittle.
Other:	Used to produce alum, potash, and alumina.

AMETHYST am'əthyst Greek *amethystos* meaning anti-intoxicant (see quartz).

ARAGONITE ərag' ə nit (Calcium carbonate) Named after the Spanish province.

Environment:	Hot springs, sedimentary formations, mollusk shells and pearls.
Chemical Formula:	$Ca\,Co_3$.
Hardness:	3.5–4.0.
Color:	Multicolored. Brown, yellowish, white, golden, clear.
Streak:	White.

Crystal System and Habit:	Orthorhombic. Prismatic, columnar, stalactitic, massive.
Cleavage and Fracture:	Poor. Conchoidal, splinter, scale.
Optical Properties:	Vitreous, greasy. Transparent, translucent.
Specific Gravity:	2.9–3.0.
Tenacity:	Brittle.
Other:	Both aragonite and calcite are calcium carbonate which is a diamorphous (having two forms) mineral. Calcite's form is trigonal while aragonite's is orthorhombic. Used as a decorative building stone. Aragonite precipitates directly from warm, usually tropical alkaline waters. Will effervesce in hydrochloric acid. See calcite.

AUGITE 'ô-jīt (calcium sodium magnesium iron titanium aluminum silicate)   Latin *augites* meaning precious stone.

Environment:	A rock-forming pyroxene group mineral present in igneous rocks such as basalt, diabase, and gabbro.
Chemical Formula:	$(Ca,Na)(Mg,Fe^{2+},-Fe^{3+},TiAl)(SiAl)_2O_6$.
Hardness:	5–6.
Color:	Black, brownish-black, dark green.
Streak:	Gray-green.
Crystal system and Habit:	Monoclinic. Massive, compact, columnar, tabular, acicular.
Cleavage and Fracture:	Good. Conchoidal.
Optical Properties:	Vitreous. Translucent, opaque.
Specific Gravity:	3.3–3.5.
Tenacity:	Brittle.

AURICHALCITE âr ə kal' sit (zinc copper carbonate hydroxide)   Greek meaning brass.

Environment:	A secondary mineral of oxidized copper-zinc deposits.
Chemical Formula:	$(Zn, Cu)_5 (CO_3)_2 (OH)_6$.
Hardness:	2.
Color:	Monochromatic. Light blue.
Streak:	Blue-green.
Crystal System and Habit:	Orthorhombic.
Cleavage and Fracture:	Perfect.
Optical Properties:	Pearly, silky. Translucent.

Specific Gravity:	3.6.
Tenacity:	Flexible.
Other:	Although composed of copper and zinc, the components of brass, it is not common enough to be a practical brass ore. Often found with malachite, azurite, and chrysocolla.

AZURITE azh' ə rit'   (copper carbonate hydroxide)   French *azur* meaning blue.

Environment:	A secondary mineral of oxidized copper deposits.
Chemical Formula:	$Cu_3(CO_3)_2(OH)_2$.
Color:	Monochromatic. Light–dark blue.
Streak:	Blue.
Crystal System and Habit:	Monoclinic. Usually prismatic or columnar.
Cleavage and Fracture:	Good. Conchoidal.
Optical Properties:	Vitreous–dull. Opaque.
Specific Gravity:	3.7–3.8.
Tenacity:	Brittle.
Other:	Azurite is usually found with malachite, the more stable of the two minerals. Azurite will easily metamorphose to malachite. Was used by some medieval artists as a cheap substitute for blue pigment instead of the more expensive lapis lazuli. Their paintings now have green skies.

BARITE bar īt   (barium sulfate)   Greek *barites* meaning weight.

Environment:	Low temperature hydrothermal veins, weathered limestone and dolomite sedimentary strata.
Chemical Formula:	$Ba[SO_4]$.
Hardness:	3–3.5.
Color:	Clear, white, yellow, golden, red, blue, brown, gray.
Streak:	White.
Crystal System and Habit:	Orthorhombic. Rhombic, bipyramidal. Tabular, columnar, massive. Often forms rosette clusters.
Cleavage and Fracture:	Perfect. Conchoidal, uneven.
Optical Properties:	Vitreous, pearly, greasy, dull. Transparent, translucent, opaque.

Specific Gravity:	4.4.
Tenacity:	Brittle.
Other:	Used commercially in the manufacture of paint pigments, cloth and paper production, and oil drilling.

CALCITE kal' sit   (calcium carbonate)   Latin *calxite* meaning burnt lime.

Environment:	A very common mineral found in all geological environments. Associated with all types of rocks.
Chemical Formula:	$CaCO_3$.
Color:	Multicolored. In Arizona, usually, clear, white, honey, and black. Worldwide, calcite is found in all colors. The mineral takes on a great variety of pretty tints and shades.
Streak:	White.
Crystal System and Habit:	Trigonal. Prismatic, rhombohedral, calenhedral. Because calcite comes in over 80 different habits, it is an excellent mineral for a specialty collection. Twinning is common.
Cleavage and Fracture:	Perfect, will cleave with little force. Conchoidal, easily pulverized.
Optical Properties:	Vitreous, pearly, dull. Transparent–opaque. Easily looses brilliance and clarity when exposed to corrosive effects of the atmosphere. Will polish in hydrochloric acid bath.
Specific Gravity:	1.8–2.0.
Tenacity:	Brittle.
Other:	Calcium carbonate is a common diamorphous mineral. It is the main constituent of a number of rocks resulting from a broad range of geologic conditions. Limestone and chalk are sedimentary. In the rock cycle, limestone then becomes the metamorphic protolith of marble. Aragonite and travertine precipitate from warm waters and hydrothermal springs. Also known as Iceland spar, satin spar, sand calcite.

CERUSSITE sër ə sit   (lead carbonate)   Latin *cerussa* meaning white lead.

Environment:	A secondary mineral of oxidized lead deposits.
Chemical Formula:	$PbCo_3$.
Hardness:	3–3.5.

Color:	Multicolored clear, white, black, smoky.
Streak:	White.
Crystal System and Habit:	Orthorhombic. Prismatic. Occurs in a wide variety of habits including columnar, tabular, acicular, druse, massive, and wheat-sheaf aggregates.
Cleavage and Fracture:	Good. Conchoidal.
Optical Properties:	Adamantine, greasy. Often reflects light better than quartz.
Specific Gravity:	6.4–6.6.
Tenacity:	Brittle.
Other:	Cerrusite is a common lead ore formed by the action of carbonated water on galena. Often found with galena and sphalerite.

CHALCANTHITE kal kan' thīt (copper sulfate pentahydrate) Greek *chaleanthon* for flowers of copper.

Environment:	Oxidized zone of sulfide copper deposits.
Chemical Formula:	$CuSO_4 \cdot 5H_2O$.
Hardness:	2.5.
Color:	Blue.
Streak:	White.
Crystal System and Habit:	Triclinic. Massive, scaly, fibrous, granular, tabular.
Cleavage and Fracture:	Indistinct. Conchoidal.
Optical Properties:	Vitreous. Translucent.
Specific Gravity:	2.2–2.3.
Tenacity:	Brittle, soluble in water, decomposes to a white dust in a dry climate.

CHALCOPYRITE kal' kɛ pi' rit (copper iron sulfide) Greek *khalkos* meaning copper.

Environment:	Igneous rocks, pegmatites, metamorphic contact zones.
Chemical Formula:	$CuFe_2$.
Hardness:	3.5–4.0.
Color:	Tarnished brass. Reflects primary colors. "Peacock ore".
Streak:	Black, greenish-black.
Crystal System and Habit:	Trigonal. Massive. Equant tetrahedral crystals with scalenohedral faces.
Cleavage and Fracture:	Poor. Uneven.

Optical Properties:	Metallic. Bright and shiny. Iridescent.
Specific Gravity:	4.1–4.3.
Tenacity:	Brittle.
Other:	Chalcopyrite is the most important of the copper ores. It is the primary mineral of porphyry cooper deposits. When weathered, it alters to secondary copper minerals such as malachite, azurite, cuprite, chrysocolla, and others. It is found in conjunction with iron pyrite and, like iron pyrite, is easily mistaken for gold.

CHRYSOCOLLA kris' ə kol' ɛ (copper aluminum hydrogen silicate hydroxide) Greek *khrusos* meaning gold and *kola* meaning glue. It was once used as a flux for soldering gold.

Environment:	A secondary mineral of oxidized copper deposits.
Chemical Formula:	$(Zn, Cu)_5 (CO_3)_2 (OH)_6$.
Hardness:	2–4. The higher the silication, the greater the hardness.
Color:	Monochromatic. Light–medium blue.
Streak:	Greenish-blue.
Crystal System and Habit:	Monoclinic. Massive, botryoidal. Often found as thin crusts and horizons.
Cleavage and Fracture:	Poor. Conchoidal.
Optical Properties:	Greasy, dull. Translucent, opaque. Vitreous if highly silicated.
Specific Gravity:	2.0–2.3.
Tenacity:	Brittle, crumbly. Will easily pulverize. May crack and become powdery if allowed to dry out.
Other:	Chrysocolla is the most common of the copper minerals. It is present at almost every copper mine and prospect in Arizona. Unfortunately, most of it is only specimen quality. Gem quality material, known as "gem silica" is highly silicated, has a hardness of 6-7, is very durable, and is a semi-precious lapidary stone. Chrysocolla will stick to a moistened finger.

CUMMINGTONITE 'kəm · iŋ · tə · nīt (magnesium iron manganese silicate hydroxide) Named for locality where first found.

Environment:	A metamorphic monoclinic amphibole. Also, a primary igneous mineral in mafic rock.
Chemical Formula:	$(Mg, Fe, Mn)_7Si_8O_{22}(OH)_2$.
Hardness:	5-6.
Color:	Dark green, grey-green, brown, light grey.
Streak:	White.
Crystal System and Habit:	Monoclinic. Fibrous. Long acicular, hair-like strands.
Cleavage and Fracture:	Good. Uneven.
Optical Properties:	Translucent, opaque. Silky.
Specific Gravity:	3.10–3.47.
Tenacity:	Brittle.

EPIDOTE 'ep-ə-dōt (Calcium iron silicate hydroxide) Greek *epididonai* meaning increase.

Environment:	Common throughout southern Arizona. Produced by alteration of rock containing base metals and metamorphism of igneous and sedimentary rock.
Chemical Formula:	$Ca_2(Al,Fe)_3SiO_{12}(OH)$.
Hardness:	6–7.
Color:	Green, yellow, gray, black.
Streak:	White.
Crystal System and Habit:	Monoclinic. Prismatic, striated, tabular, acicular, massive, granular, fibrous, lamellar. Over 200 different forms.
Cleavage and Fracture:	Perfect. Conchoidal.
Optical Properties:	Opaque. Vitreous, pearly, resinous.
Specific Gravity:	3.3–3.5.
Tenacity:	Splintery.
Other:	Fluoresces a weak red.

FLUORITE floor' it (Calcium fluoride) Latin *fluere* meaning to flow.

Environment:	Ore veins, sedimentary rocks, plutonic igneous rocks.
Chemical Formula:	CaF_2.
Hardness:	4.
Color:	Multicolor. Fluorite comes in a delightful array of brilliant colors and subtle hues. In Arizona, the most common colors are clear, green, and purple shades. Fluorite colors will often fade in sunlight.

Streak:	White.
Crystal System and Habit:	Cubic. Octahedron. Usually found in compact massive veins. Individual cubes or, more rarely, octahedron crystals may form on massive vein surfaces.
Cleavage and Fracture:	Perfect. Conchoidal, splintery, uneven. Cleaves easily but not in all directions. Do not expect a single tap to precipitate a handful of perfect little octahedrons.
Optical Properties:	Vitreous. Transparent–translucent. Very gemmy.
Specific Gravity:	3.0–3.2.
Tenacity:	Brittle.
Other:	The term fluorescence is derived from the name of this mineral.

GALENA gə lë nɛ   (Lead sulfide)   Latin *galena* meaning lead ore.

Environment:	Igneous rock, sedimentary strata, low-medium temperature lead ore veins.
Chemical Formula:	PbS.
Hardness:	2.5.
Color:	Lead-gray, silvery, bluish.
Streak:	Lead-gray.
Crystal System and Habit:	Cubic. Occurs as individual cubes, octahedron crystals, and compact masses.
Cleavage and Fracture:	Perfect. Even fracture.
Optical Properties:	Metallic. Opaque. Looks like highly polished metal. May display bluish iridescence.
Specific Gravity:	7.2–7.6.
Tenacity:	Brittle.
Other:	An important lead and silver ore. Often found with other crystals such as fluorite, calcite, sphalerite, pyrite, and others.

GARNET gär' nit   (aluminum and calcium silicate)   Latin *graniticus* meaning like a pomegranate seed.

Environment:	Metamorphic and igneous rock.
Chemical Formula:	$Fe_3Al_2(SiO_4)_3$.
Hardness:	6.5–7.5.
Color:	Multi*Color:* Yellow, orange, green, brown, red, violet, black.
Streak:	White.

Crystal System and Habit:	Cubic. Dodecahedral, rhombdodecahedral, icositetrahedral crystals. Globular, granular, and massive habits.
Cleavage and Fracture:	Poor. Conchoidal, uneven, splintery.
Optical Properties:	Vitreous–greasy. Translucent–opaque.
Specific Gravity:	3.5–4.3.
Tenacity:	Brittle. Massive boulders very hard to break.
Other:	There are several types of garnets divided into two categories. Pryralaspite contains the pyrope, almandine, and spessartine types. Ugrandite contains the uvarovite, grossular, and andradite varieties. Massive garnet deposits are ground up for use as abrasives (sandpaper) and crystals are faceted into semi-precious gem stones.

HEMATITE hë mə tit' (iron oxide)   Greek *haimatites* meaning blood-like stone.

Environment:	A common mineral found in sedimentary strata and igneous flows.
Chemical Formula:	Fe_2O_3.
Hardness:	5.5–6.5.
Color:	Black, reddish, rusty.
Streak:	Reddish-brown.
Crystal System and Habit:	Trigonal. Tabular. Granular, massive, botryoidal, flat slabs.
Cleavage and Fracture:	Poor. Conchoidal and uneven fracture.
Optical Properties:	Metallic–earthy. Some is very shiny like well-tooled steel. Some is dull, dirty, and rusty looking. The specularite variety is beautifully iridescent. Granular druses are highly vitreous.
Specific Gravity:	5.2–5.3.
Tenacity:	Brittle, crumbly.
Other:	Hematite is the basic iron ore. Crystals are faceted into semi-precious gem stones. Often an unwelcome mineral that stains and discolors other crystals.

MALACHITE mal' ə kit' (copper carbonate hydroxide)   Greek *molokhitis* meaning mallow-green stone.

Environment:	A secondary mineral of oxidized copper deposits.

Chemical Formula:	$Cu_2(CO_3)(OH)_2$
Hardness:	3.5–4.5. 6 if highly sicliceous.
Color:	Monochromatic. Various shades of green. Often banded.
Streak:	Light green.
Crystal System and Habit:	Monoclinic. Botryoidal, stalactitic. Crystals are rare and usually prismatic needles in colonies and fibrous aggregates.
Cleavage and Fracture:	Good. Conchoidal, splintery. Banded botryoidal veins may tend to exfoliate along bands.
Optical Properties:	Vitreous, silky. Opaque. Will take a bright gemmy shine if highly siliceous.
Specific Gravity:	4.0.
Tenacity:	Brittle.
Other:	Gem quality malachite is a popular gemstone and lapidary material. Russia produced veins large enough to fashion into large statuary and decorative carvings. Zaire produces excellent African animal malachite statuary. And, malachite jewelry has worldwide appeal.

MARBLE mär bəl   Greek *Maramos* meaning white stone. See calcite.

MIMETITE mim' i tit'   (lead arsenic chloride)   Greek *mimes* meaning imitator.

Environment:	A secondary mineral formed in lead oxidation zones.
Chemical Formula:	$Pb_3[Cl(AsO_4)_3]$.
Hardness:	3.5.
Color:	Deep, rich orange-yellow. White, greenish, gray.
Streak:	White.
Crystal System and Habit:	Hexagonal. Botryoidal, globular, granular, acicular. Forms delicate encrustations.
Cleavage and Fracture:	Poor. Conchoidal.
Optical Properties:	Resinous. Translucent. The best quality mimetite presents a bright, intense appearance.
Specific Gravity:	7.1.
Tenacity:	Brittle.
Other:	So named because it resembles, imitates, pyromorphyte. A member of the apatite group. Makes a stunning background mineral for wulfenite crystals such as the specimens found

at the Rowley Mine in Arizona and the San Francisco Mine in Sonora, Mexico.

MUSCOVITE (MICA) 'məs-kə-vīt (Potassium aluminum silicate hydroxide) Named after Muscovy Principality, Russia.

Environment:	Occurs in a wide variety of environments: granitic pegmatites, schist, gneiss, detrital sediments.
Chemical Formula:	$KAL_2(Si_3AL)O_{10}$.
Hardness:	2.0–2.5.
Color:	Colorless, green, yellow, red, brown, violet.
Streak:	White.
Crystal System and Habit:	Monoclinic. Massive, globular, scaly, tabular, lamellar.
Cleavage and Fracture:	Perfect. Exfoliates in thin laminates like pages in a book.
Optical Properties:	Pearly, silky, vitreous. Transparent, translucent.
Specific Gravity:	2.8–3.0.
Tenacity:	Flexible and elastic.
Other:	Greek name is *phengites* meaning transparent stone used for windows. "Isinglass" used in stove door windows. Also used in electrical insulation, lubricants, and fireproofing.

ONYX on' iks Greek *onix* alluding to the color of the fingernail. See calcite.

OPAL ö' pəl (hydrous silica) Sanskrit *upala* meaning precious stone.

Environment:	Sedimentary strata, volcanic rocks.
Chemical Formula:	$SiO_2 \cdot mH_2O$.
Hardness:	5.5–6.5.
Color:	White, gray, yellow, orange, red, green, blue, violet, black.
Streak:	White.
Crystal System and Habit:	Amphorous. Botryoidal, massive, stalactitic, rock forming in cavities and veins. Will replace organic tissue to form fossils, especially wood.
Cleavage and Fracture:	Does not cleave. Fractures in sharp shards and conchoidal patterns like glass.
Optical Properties:	Vitreous, pearly, greasy, dull. Transparent, translucent, opaque. Gem quality fire opal is adamantine.
Specific Gravity:	2.5–2.6.

Tenacity:	Brittle. Will craze and crack apart if allowed to dry out.
Other:	Most opal found in Arizona is white, opaque, and resembles porcelain. It is usually found in geodes and occasionally in fossils. There is a least one privately owned fire opal mine in Arizona.

PYRITE (IRON PYRITE) pi' rit   (iron disulfide)   Greek *pyrites* meaning of fire.

Environment:	All geological environments.
Chemical Formula:	6.0–6.5.
Hardness:	FeS_2.
Color:	Gold, silvery, brassy. Called "fool's gold."
Streak:	Greenish-black.
Crystal System and Habit:	Cubic. Found as well-defined, individual cubes and cubic clusters as well as massive and granular forms. Crystal faces are usually striated.
Cleavage and Fracture:	Poor, indistinct. Conchoidal, uneven.
Optical Properties:	Metallic, iridescent. Opaque. Oxidation will cause a variety of colors similar to oil on water.
Specific Gravity:	4.8–5.0.
Other:	Iron is the most common of many metal disulfides that share the pyrite structure. Others include copper, lead, manganese and platinum. Weathered iron pyrite is the cause of the ugly yellow stained ground common to many abandoned mine sites throughout Arizona.

QUARTZ kwô rts   (silicon oxide)   Greek *krystallos* meaning ice.

Environment:	All geologic environments.
Chemical Formula:	SiO_2.
Hardness:	7.
Color:	Clear, white (milky), yellow (citrine), green (aventurine), apple-green (chrysoprase), blue-violet (amethyst), pink (rose quartz), and orange-red (carnelian).
Streak:	White.
Crystal System and Habit:	Trigonal. Massive, cryptocrystalline, tabular, columnar, druse, acicular, striated, crusty, concretions, twining. Sizes and shapes vary

	greatly. Some crystals include phantom shapes, aesthetic inclusions, and other minerals.
Cleavage and Fracture:	Does not cleave. Conchoidal.
Optical Properties:	Vitreous–dull. Transparent–opaque. Highly vitreous and transparent quartz, especially the colored varieties, are used for gemstone faceting and lapidary.
Specific Gravity:	2.6.
Other:	Because it is so common and comes in so many varieties, quartz is probably the most commonly collected mineral in Arizona. Varieties include agate, fire agate, jasper, chert, flint, and others. Quartz crystals and chalcedony, cryptocrystalline quartz composed of fibrous micro textures in botryiodal and irregular masses, are Arizona rockhound favorites.

SCHEELITE shā lit (calcium tungstate) Named after Karl W. Scheele, discoverer of tungsten.

Environment:	Contact metamorphic deposits, hydrothermal veins, pegmatites, placer deposits. Usually found with wolframite.
Chemical Formula:	$CaWO_4$.
Hardness:	4.5–5.0.
Color:	Clear, white, gray, yellow, orange, green, red.
Streak:	White.
Crystal Streak and Habit:	Tetragonal. Octahedral, massive, columnar, forms crusts.
Cleavage and Fracture:	Good resembling octahedron. Uneven, conchoidal.
Optical Properties:	Adamantine. Transparent, translucent. Fluoresces pale blue.
Specific Gravity:	6.1–6.2.
Tenacity:	Brittle.
Other:	Scheelite is an important tungsten ore widely distributed in Arizona.

SELENITE sel ə nīt (calcium sulfate hydrate) Greek *selenites* meaning stone of the moon. A variety of gypsum. Also called satin spar.

Environment:	Sedimentary rocks, salt deposits, oxidation zone of sulfur ore. An evaporate found in Permian and Triassic marine sediments.

Chemical Formula:	Ca[SO$_4$] . 2H$_2$O.
Hardness:	1.5–2.0.
Color:	Clear, yellowish, brownish.
Streak:	White.
Crystal System and Habit:	Monoclinic. Prismatic. Acicular, fibrous, micaeous. Warps, curves, forms rosettes, concretions, and massive plates.
Cleavage and Fracture:	Very good. Splintery, easily pulverized.
Optical Properties:	Vitreous, silky, pearly. Transparent, translucent.
Specific Gravity:	2.3–2.4.
Tenacity:	Flexible.
Other:	Name derives from the ancient belief that stone waxed and waned like phases of the moon. Used for making plasterboard. Gypsum is Greek *gypsos* for plaster.

SERPENTINE sər pən tēn   (magnesium silicate hydroxide)   Latin *serpentinus* meaning like the movement of a snake.

Environment:	Formed by magnesium limestone contact metamorphism and in hydrothermal veins. A very pervasive mineral easily replacing a host of dark colored magnesium-silicate rocks. Whole mountains along California's coast are Serpentine.
Chemical Formula:	Mg$_3$ Si$_2$O$_5$(OH)$_4$.
Hardness:	3–4.
Color:	Greenish-yellow, brownish-red, bluish-violet, black.
Streak:	White, grey.
Crystal System and Habit:	Monoclinic. Massive, wheat-sheaf, fibrous, microcrystalline.
Cleavage and Fracture:	Good (chrysotile form). Splintery.
Optical Properties:	Silky. Shimmers and shines. Translucent, opaque.
Specific Gravity:	3.1–3.2.
Tenacity:	Elastic, sectile. Fibers will separate and fluff when teased.
Other:	Serpentine is a secondary mineral formed from the alteration of such minerals as amphibole, olivine, pyroxene, peridoite, and other igneous rocks. In Arizona, the chrysotile form of

serpentine was mined for the production of asbestos products. Chrysotile is safer than crocidolite, the other asbestos producing type of serpentine. Serpentine feels and looks like plastic and can be used as a sculpting and lapidary medium.

SIDERITE sid ə rīt (Iron carbonate) Greek *sideros* meaning iron.

Environment:	Found in sedimentary rocks and as a metasomatic mineral formation in hydrothermal veins.
Chemical Formula:	$FeCO_3$.
Hardness:	3.5–4.
Color:	Brownish-black, yellowish, grey, white. Brown color is usually a thin, oxidized, outer layer.
Streak:	White, yellowish.
Crystal System and Habit:	Trigonal. Rhombohedral, scalenohedral. Crystals are often curved and distorted.
Cleavage and Fracture:	Perfect. Conchoidal.
Optical Properties:	Vitreous, pearly. Translucent, opaque.
Specific Gravity:	3.7–3.9.
Tenacity:	Brittle.
Other:	Siderite is an important iron ore. It is a member of the calcite family. It forms when hot gases and solutions escape through cracks in sedimentary rock dissolving it and forming new secondary minerals.

TRAVERTINE trav' ɛr tën Latin *lapis Tiburnius* meaning stone of Tibur. See Calcite.

VANADINITE və nad' n it' (lead vanadate chloride) Old Norse *vanada*, the Scandinavian god of fertility.

Environment:	A secondary mineral found in oxidized lead deposits. Often found with wulfenite, cerussite, and descloisite.
Chemical Formula:	$Pb_5 (VO_4)_3 Cl$.
Hardness:	3.
Color:	Deep ruby-red, orange, brownish, yellowish.
Streak:	Yellow-brown.
Crystal System and Habit:	Hexagonal. Dipyramidal, columnar, barrel shaped, reniform, botryoidal, acicular, massive.

	Forms crusts and druses.
Cleavage and Fracture:	Does not cleave. Conchoidal.
Optical Properties:	Adamantine, greasy. Translucent, opaque. Can be very gemmy.
Specific Gravity:	7.2.
Tenacity:	Brittle.
Other:	The source of vanadium that is used to harden steel and as a minor source of lead. A rather rare mineral, its discovery in Arizona in the 19th century caused considerable interest among eastern mining speculators and developers.

WILLEMITE 'wil-ə mīt   zinc silicate.   Named for William I (1813–1840) King of The Netherlands.

Environment:	Oxidized zones of zinc deposits, schists, granitic pegmatites.
Chemical Formula:	Zn_2SiO_4.
Hardness:	5.5.
Color:	Colorless, green, yellow, brown, blue-violet.
Streak:	White.
Crystal System and Habit:	Trigonal.
Cleavage and Fracture:	Good. Conchoidal.
Optical Properties:	Greasy, vitreous. Transparent, translucent.
Specific Gravity:	4.0–4.2.
Tenacity:	Brittle, splintery.

WULFENITE   wool' fə nit'   (lead molydate)   German *wulfenite* named after Franz X. von Wulfen (1728–1805), an Austrian mineralogist.

Environment:	A secondary mineral found in oxidized lead deposits. Often found with mimetite in Arizona.
Chemical Formula:	$Pb\,Mo\,O_4$.
Hardness:	3.
Color:	Yellow, orange, red. Colors can be bright and intense.
Streak:	White.
Crystal System and Habit:	Tetragonal. Pyramidal, basil pinacoid, tabular, short columnar, acicular, massive. Forms crusts and druses.
Cleavage and Fracture:	Does not cleave. Uneven, conchoidal.

Optical Properties:	Adamantine, greasy. Transparent–opaque. Can be very brilliant and gemmy.
Specific Gravity:	6.5–6.9.
Tenacity:	Brittle. Easily pulverized.
Other:	Perhaps Arizona's favorite mineral among collectors. The "Wulfenite Project" sponsored by the Arizona Mining and Mineral Museum Foundation has so far identified over 200 wulfenite bearing locations within the state.

Site Difficulty Scale

Difficulty Code	Road Leading to Site	Terrain at Site	Extraction of Material
Easy	*OK for passenger cars.*	*Minimal effort and experience required.*	
1	Paved road	Flat, open, smooth ground	Light, visible, loose float
2	Maintained gravel	Flat, scattered vegetation, small rocks	Small, medium-sized, half buried float
3	Unmaintained gravel, good condition	Gentle slopes, light vegetation, fairly rocky	Small, medium-sized pieces visible in tailings or rubble
Moderate	*High-clearance vehicle required.*	*Average strength and skill required.*	
4	Unmaintained gravel, fair condition	Moderate slopes, vegetation, large rocks, ledges	Light hand tools helpful to remove pieces from ground
5	Bumpy, rocky, sandy, shallow ruts	Tailings piles, open trenches and shafts, obstacles	Hand tools required to remove material from soil or matrix
6	Large rocks, deep ruts and erosions, steep hills	Rough, steep, slippery slopes	Moderate digging, prying, hammering required
Hard	*Four-wheel drive only.*	*Superior stamina and expertise required.*	
7	Steep, heavily eroded, sharp rocks, exposed ledges	Very steep, heavily vegetated rocky cliffs, very hard to climb	Extensive digging, required to extract material
8	Dangerous, steep, mountainous, no turnarounds, narrow, scary	Unstable, deep excavations, decayed structures, loose tailings slides	Exceptionally hard rocks requiring heavy hammering
9	Most treacherous, low gear, 4-wheel, no stopping, no backing up	Exhausting, requires rigorous climbing	Material is exceptionally difficult to obtain due to its location

At the beginning of each chapter, a three-number code appears corresponding to the descriptions contained in this site difficulty scale. The first number represents the road conditions, the second represents the site terrain conditions, and the third represents the difficulty of collecting the material.

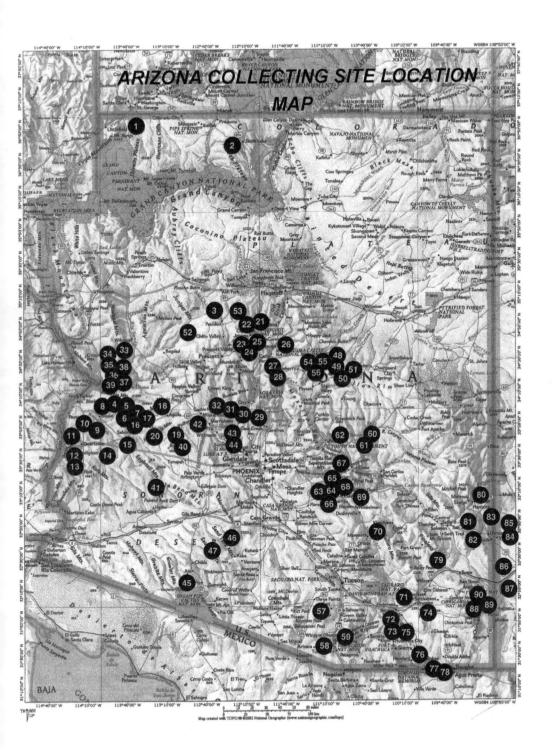

Minerals

THIS BOOK IS WRITTEN for the hobbyist, not the scientist. However, a basic understanding and knowledge of the scientific principles of mineralogy adds more meaning to and appreciation of the art of collecting. It is usually not necessary to subject your rock and mineral finds to sophisticated and expensive laboratory analysis to determine what they are. Being able to recognize the basic characteristics of rocks and minerals, knowing what differentiates one from another, and knowing how to conduct simple field tests will enable you to identify most of your discoveries fairly accurately.

Mineral Identification

The best and easiest way to field-test a mineral is also the most obvious. What does it look like? Consider its color, luster, and transparency. Minerals, like most everything else in nature, come in all colors and color combinations. Some are even named after their color such as azurite. If you rub a piece of azurite across a rough unglazed porcelain plate, it will leave a blue mark. This procedure is called a streak test. However, a piece of blue fluorite will leave a white streak. The difference in streak is because, as a general rule, the coloration of a monochromatic mineral, such as azurite, is a constant identifying characteristic of that mineral. Some azurite may be darker or lighter blue than others, but all azurite streaks blue nevertheless. Multicolored minerals, such as fluoride, cannot be identified by the color you see because the apparent color is caused by trace amounts of a foreign substance that is contaminating the true color of the mineral. Minerals streak white, black, brown, green, yellow, red, and blue. See Table 1.

Along with color, luster is one of the mineral properties that first catches our eye. Luster, or shininess, depends upon the condition of a mineral's reflecting surface. When a ray of sunshine bounces off a smooth glassy crystal surface hidden away among the dull drab country rock, it gets our attention. The more light the crystal surface reflects, the brighter the shine. The more it absorbs or diffuses light, the duller it appears. Brilliant luster is the hallmark of fine gem and lapidary stone. See Table 2.

A mineral's optical properties, how light waves pass through it, are also clues to its identity. Some minerals are opaque. They appear to be solid because no light can pass through them. Others are translucent somewhat like frosted

TABLE 1
Color Streak

Monochromatic Minerals

Color Streak	Visible Color	Minerals
Black	Gray, black, metallic tarnishes	Arsenopyrite, bixbyite, bornite, chalcopyrite, cobaltite, galena, graphite, hematite, magnetite, marcasite, pyrite, stibnite
Blue	Blue	Azurite, aurichalcite, covellite, cynotrichite, linarite, vivianite
Brown	Brown, blackish-brown, rusty color	Chromite, franklinite, goethite, limonite, rutile, sphalerite
Green	Green, blue-green	Atacamite, actinolite, brochantite, chlorite, chrysocolla, conichalcite, dioptase, malachite, rosasite
Red	Red, orange-red	Cinnabar, chalcotricite, copper, cuprite, hematite
Yellow	Yellow, yellow-green, yellow-orange	Gold, crocoite, orpiment, realgar, siderite, sulfur, vanadinite

Multicolor Minerals

White	Colorless, white, yellow, orange, red, violet, blue, green, brown, gray, black, and a wide range of tints, shades, and color combinations. Not all minerals come in all colors.	Alunite, aragonite, autinite, barite, beryl, calcite, celestite, cerussite, corundum, dolomite, fluorite, garnet, gypsum, halite, hemimorphite, kyanite, mica, mimetite, natrolite, olivine, opal, quartz, scheelite, selenite, serpentine, smithsonite, spodumene, topaz, tourmaline, willemite, wulfenite

glass. The most striking are the transparent minerals especially the crystals that, being free from deformities and contaminates, are indeed "crystal" clear. Some minerals, such as ulexite, are birefringent meaning that the mineral splits the light wave in two. One wave travels through the mineral in a straight line while the other is bent causing distorted or displaced images. A few minerals, such as labradorite, are pleochroic meaning they show different colors when viewed from different angles.

Crystals are the flowers of the mineral kingdom. Being able to determine a mineral's crystal structure provides a major clue to its identity. The crystal

TABLE 2
Luster

Luster	Description	Minerals
Adamantine	Brilliant, sparkly	Found in transparent highly refractive minerals. Anglesite, cerussite, diamond, sagenite, scheelite, sphalerite, sulfur, wulfenite
Chatoyant	Changing luster or color	Reflects narrow band of light like a cat's eye. Labradorite, smithsonite
Earthy	Dull, drab	Characteristic of minerals that refract light poorly. Basalt, chalk, sandstone, limestone, graphite, Volcanic tuff
Greasy	Somewhat dulled by microscopic impurities	Amber, aragonite, barite, cerussite, chromite, common opal, olivine, smithsonite, vanadinite
Metallic	Opaque, highly reflective, shiny	Chalcopyrite, galena, gold, hematite, magnetite, platinum, pyrite, silver, copper, nickel
Pearly	Glows more than shines	Occurs in minerals made up of fine layers or parallel surfaces. Alunite, anhydrite, aurichalcite, barite, calcite, dolomite, kyanite, orthoclase, siderite
Silky	Shimmery, a wavering sheen or glimmer	Characteristic of fibrous minerals. Asbestos, chalcedony, geothite, gypsum, limonite, malachite, natrolite
Vitreous	Shiny like glass	Typical of many mineral crystals. Azurite, adamite, dioptase, emerald, epidote, fluorite, garnet, quartz, rhodochrosite, selenite

structures depicted in Table 3 are idealized representations. Actually, such flawless specimens are rare and, therefore, prized by collectors. In the real, imperfect world, contamination, geologic forces, competition for space, and other natural influences adversely affect the crystallization process resulting in impure and malformed crystals. Sometimes, these imperfections are desirable as in the case of contaminate minerals, such as rutile, being encased in quartz crystals. A few minerals, the amphorus type such as opal and obsidian, have no

TABLE 3
The Crystal Systems

Model	Crystal System	Axes	Crystal Forms
	Isometric or cubic	Three equal length axes intersecting at right angles	There are 15 different types of forms in the isometric system. Types have 4, 6, 8, 12, 24, and 48 faces. The most common are cubes and octahedrons
	Tetragonal (six sided)	Three axes intersecting at right angles with one longer or shorter than the others	There are 33 different types of forms among the remaining 6, non-isometric, systems
	Hexagonal (six sided)	Three equal length axes in a horizontal plain intersecting at 120° with a fourth axis intersecting at 90°	Types have 1, 2, 3, 4, 6, 8, 12, 16 and 24 faces. Types are monohedron -1 face, pinacoid–2 faces, dome–2 faces, sphenoid–2 faces, 7 prisms (3,4,6,8,12 faces), 7 pyramids (3,4,6,8,12 faces)
	Trigonal (three sided)	Same as hexagonal except c axis has threefold symmetry	7 dipyramids (3,4,6,8,12 faces)
	Orthorhombic (flat-shaped)	Three different length axes intersecting at 90°	3 rapezohedrons (3,6,12 faces), 2 scalenohedrons (8 or 12 faces)
	Monoclinic (singly inclined)	Three axes with two intersecting at 90° and one intersecting at more than 90°	Rhombohedron-2 faces, rhombic disphenoid- 4 faces, tetragonal disphenoid- 4 faces
	Triclinic (thrice inclined)	Three different length axes with none intersecting at 90°	

TABLE 4
Crystal Habits

Crystal Habit	Mineral	Description	Plate and Photograph
Acicular	Natrolite, okenite, stibnite	Needlelike	19-3
Bladed	Hematite, kyanite, gypsum	Like a knife blade	12-2, 14-2
Capillary or filiform	Cummingtonite	A hairlike mass	12-3
Colloform (globular, reniform, botryoidal, mammillary)	Malachite, hematite, chalcedony	Similar to a bunch of grapes, curvaceous	14-3, 19-4
Columnar	Tourmaline, beryl	Elongated, postlike	19-2
Concretionary	Sedimentary rock	Rocks cemented together	—
Coralloid	Aragonite (flos ferré)	Branches like coral.	—
Dendritic	Iron or manganese oxide deposits	Tree-like. Flat, two-dimensional	20-6
Equant	Garnet	Equal diameter in all directions like a soccer ball	—
Lamellar	Graphite, mica, lepidolite	A mass of layers	16-3
Prismatic	Tourmaline, beryl	Elongated like a tube	—
Striated	Pyrite, quartz	Parallel grooves or scratches on crystal faces	—
Twinning	Quartz, calcite, fluorite	Grown together in specific ways	—
Druse	Quartz, fluorite, calcite, and several more minerals	A wash or coating of small tightly packed crystals	20-1, 20-5
Tabular	Wulfenite, angel-wing calcite, quartz	Thin flat square or rectangular plate	16-1
Fibrous	Aurichalcite, asbestos, gypsum	Tightly banded strands. Fluffs like cotton when teased	18-3, 18-5

TABLE 4 (cont.)
Crystal Habits

Crystal Habit	Mineral	Description	Plate and Photograph
Massive	Most all minerals	Large, chunky, cryptocrystalline.	5-1, 17-3
Stalactitic	Calcite dripstone, malachite	Elongated icicle-shaped formations	19-4
Wheat-sheaf	Epidote	Similar to a bundle of straw	13-3
Micaeous	Muscovite, mica, lepidolite, selenite	Thin flat sheets like the pages of a book	12-1

crystalline structure. They look like mineral jelly or lumps of glass. Often, minerals, quartz and calcite for example, will exist in several different crystal forms called habits. The crystal shape varies, but the crystal system remains the same. For example, some quartz crystals may be long and skinny, others may be short and squatty, and some may be flat and wide. But, they all are trigonal having six faces. Calcite sports 80 or more habits. See Table 4.

If a mineral's appearance is insufficient to reveal its identity, then the way it reacts to external forces applied to it will help identify it. You may have to get physical with it. A mineral's hardness, toughness, and fracture pattern can easily be tested in the field or workshop. Hardness is measured by scratching or cutting. A scratch test is used to gauge hardness on a scale of 1–10. Developed by mineralogist Frederich Mohs in 1812, this scale lists ten minerals arranged from the softest (talc) to the hardest (diamond). Each mineral on the scale can scratch the one preceding it and be scratched by the one following it. However, the difference in hardness between minerals is not equal. Diamonds are much more than ten times harder than talc. The cutting scale developed by petrologist August Rosiwal (1860-1923) gives a clearer idea of the relative hardness differences between the Mohs scale minerals. See Table 5.

Minerals are characterized by the way they bend or break apart. Minerals that separate in a random, unpredictable, irregular, and unpredictable pattern are said to fracture. Fracture occurs in unpredictable ways. See Table 6. Minerals that separate along crystal faces within their internal structure are said to cleave. Cleavage splits minerals along crystal planes, fracture breaks across them. Most, but not all, minerals will both cleave and fracture. Tenacity or toughness of a mineral is determined by its behavior when scratched or bent. Tenacity is a measure of the strength of its chemical bonding. When pressured to bend, minerals will display different degrees of brittleness that can be tested by scratching. If the scratch furrow powder remains next to the scratch without

TABLE 5
The Mohs Hardness and the Rosiwal Cutting Scales

Mohs Hardness Scale	Mineral	Scratch Instrument	Rosiwal Cutting Scale
1	Talc	Fingernail	0.03
2	Gypsum	Copper Wire	1.25
3	Calcite	Dolomite	4.5
4	Fluorite	Scheelite	5.0
5	Apatite	Common Opal	6.5
6	Feldspar	Steel nail, pyrite	37.0
7	Quartz	Danburite, schorl	120.0
8	Topaz	Carbide drill bit	175.0
9	Corundum	Diamond drill bit	1000.0
10	Diamond	Diamond	140000.0

falling away, as with copper, the mineral will be flexible. If the powder easily drops or blows away, as with calcite, then it is brittle. If the scratch furrow leaves no dust, then the mineral is capable of being cut with a knife. Some minerals are flexible and elastic. When bent, they will return to their original shape when the pressure is relieved. Others will bend and retain their new shape. See Table 7.

Density is another useful identifying characteristic of minerals. Specific gravity measures density by comparing the weight of a mineral to a like amount of water. Water has a specific gravity value of 1. Most minerals have a specific gravity of between 2 and 3. Some minerals such as barite, scheelite, ore minerals,

TABLE 6
Fracture Patterns

Type	Mineral	Description
Conchoidal	Quartz	The fracture occurs in a random, curved, concave or convex manner resembling the shape of a sea shell.
Earthy	Talc	The fracture results in a soft irregular surface. Chalky.
Even	Jasper	Fractures into a smooth, nearly flat surface.
Hackly	Hematite	Fractures in sharp irregular angles like broken cast iron.
Uneven	Pyrite	Fractures in a rough and irregular manner.

TABLE 7
Tenacity

Characteristic	Mineral	Description
Brittleness	Most minerals are brittle.	Will break into pieces when struck or cutting is attempted.
Elasticity	Mica	Will easily bend, but will resume original shape when pressure is released.
Flexibility	Gypsum	Will stay in bent position when pressure is released.
Ductility	Native metals, especially copper.	Can be stretched into wire.
Malleability	Native elements, especially gold.	Can be beaten and pressed into thin sheets such as gold leaf.
Sectility	Chalcocite	Can be cut with a knife. Most minerals are not sectile.

and native metals are much heavier. For simple field-testing, hold a piece of mineral of known density in one hand and an equal size piece of the mineral you wish to identify in the other and compare the two. This may be a rough and imprecise identification test, but it will quickly narrow the field of possibilities. See Table 8 for specific gravity comparisons of certain well-known minerals.

TABLE 8
Density Comparison of Some Well Known Minerals

Mineral	Specific Gravity
Water	1.0
Chrysocolla	2.0
Calcite	2.6
Quartz	2.65
Fluorite	3.2
Malachite	4.0
Pyrite	5.0
Hematite	5.3
Galena	7.6
Bismuth	9.8
Mercury	13.8

Mineral Environments

Haphazard prospecting for collectable minerals may be fun and even prove successful sometimes. But, knowing which geologic formations are most likely to contain the right stuff is more productive. The earth is composed of igneous, sedimentary, and metamorphic rocks. Each of these three types is host to different types of minerals. Arizona has excellent exposures of each type available for your prospecting pleasure (see map of Arizona Rocks, page 227).

Each type of rock is created by interaction with the other two in an ongoing process of transformation called the rock cycle. Igneous rock forms when molten magma approaches or penetrates the Earth's surface, cools, and crystallizes. Once on the surface, igneous rock is gradually broken down by weathering, erosion, and tectonic forces into smaller and smaller fragments and grains. Wind, water, and gravity transport this material and deposit it into horizontal sediments. Mountain rock wears down filling the valleys with gravel, sand, dirt, and clay. Waves wear away coastlines and deposit layer upon layer of sand and silt on the ocean bottom. As sediments continue to pile up, increased pressure and temperature compact and cement the accumulated material together fusing it into a new type rock—sedimentary rock. As time passes, the increasing weight of the thickening sedimentary stratum forces it to sink. The tremendous pressure and temperature generated by this process physically and chemically changes and recrystallizes the deeply buried sedimentary rock thereby transforming it into metamorphic rock. As the rock cycle continues, the newly formed metamorphic rock will eventually succumb to even more heat and pressure, be transformed into igneous rock, and the rock cycle process begins all over again.

Igneous Rock (Latin *igneus* for fire)

At the surface of the earth, there are two categories of igneous rocks. Those that are deposited on the surface by volcanic eruption are known as extrusive igneous rocks. Extrusive rocks are evident in the San Francisco Mountains near Flagstaff, the lava flows along U.S. Route 89 between Flagstaff and Cameron, the road cuts along Interstate 17 between Black Canyon City and the Verde Valley, and at hundreds of other locations throughout Arizona. Magma that rises and cools just below the surface is known as intrusive igneous rock. It forces its way upward from the earth's interior intruding into the rock above it. The cooled intrusions are called plutons after Pluto the god of the underworld. Plutons exposed by erosion are visible on U.S. Route 93 near a place called Nothing and along State Route 87 at the western base of the Mazatzal Mountains. As igneous rocks begin to cool, minerals contained in the advancing magma tend to concentrate and are forced under great pressure into voids, openings, and cracks in the rocks above. Some minerals such as fluorine and chlorine along with water may become vaporized by the magma's extreme heat. The resulting gas bubbles leave egg shaped cavities in the hardened igneous rock. These cavities may become the home of minerals that crystallize out of mineral laden water that drains into them from above forming geodes and crystal lined vugs. Excellent examples of these are found at the Maggie Mine, site 38 (see page 336). See Table 9 for the mineralogy of igneous rock types.

Granite pegmatites (Greek *pegma* for "fasten together") are igneous formations that deserve special mention for they are the "mother of crystals". Pegmatites are light colored intrusive veins of course granite rock composed primarily of feldspar, quartz, and mica. Pegmatites are the final stage of the cooling and crystallizing process, the last gasp, where all of the concentrated minerals and gasses fasten together to create nature's most spectacular mineral specimens. Pegmatites worldwide have yielded feldspar and quartz crystals weighing several tons. In America, feldspar, spodumene, and beryl crystals up to 40 inches long have been discovered in New England and the Dakotas. Be alert for pegmatites. If you are lucky, you may discover one containing good quality schorl, beryl, tourmaline, and other large crystal collectables. Make my day.

Igneous rock formations, along with sedimentary and metamorphic formations, also contain mineral deposits caused by hydrothermal solutions. Under immense pressure, superheated water containing dissolved gasses form highly corrosive acids, such as hydrofluoric and hydrochloric, which dissolve a host of minerals which are then concentrated, transported, and redeposited. At the redeposit location, mineral replacement and oxidation processes may result in rich mineral and ore deposits containing a host of crystals and pseudomorphs that form in the oxidation zone.

Below the earth's surface, igneous rocks, which make up 95% of the earth's volume, remain mineralogically stable. It is when they come into contact with atmospheric gases and water at the surface that they come alive, so to speak, giving birth through oxidation and other chemical reactions to the gorgeous minerals and crystals we love to collect. That same atmosphere also provides the erosive forces of wind and water that wear down igneous formations revealing their hidden gems.

Sedimentary Rock (Latin *sedimentum* for settle)

Sedimentary rocks are best known as the final resting place for fossils. However, some of the most desirable gem quality rocks and minerals sought by collectors are produced by chemical weathering in sedimentary oxidation zones of metal ore bodies. In arid Arizona, the oxidation zone, that portion of the ore body that is above the water table, is often very deep allowing mineral laden surface water containing dissolved oxygen to percolate far below the surface through loosely compacted breccias, conglomerates, and various strata. As the water penetrates

TABLE 9
Igneous Rock Mineralogy

Intrusive — Plutonic

Rock Type	Landscape Form	Description	Collectable Minerals
Granite – Granodiorite	Rounded, exfoliating domes	Light color. Composed of quartz, feldspar mica. Uniform texture	Apatite, beryl, fluorite, gold, lead, silver, tin, zinc, zircon
Syentite – Monzonite	Rounded domes	Light color. Composed of feldspar and orthoclase. Uniform texture like granite	Apatite, corundum, garnet, olivine, sodalite, pyrite, zircon
Diorite	Round hills, weathered outcrops	Dark color. Composed of feldspar and quartz. Granular, crumbly	Apatite, magnetite, rutile
Peridotite	Rounded hills	Dark color. Composed of olivine and pyroxene. Granular, crumbly	Chromite, diamond, pyrope garnet, peridot, serpentine, soapstone
Granite – Pegmatite	Dikes, stringers, veins cutting through other formations	Light color. Rough, course texture. "The mother of crystals"	Aquamarine, aventurine, beryl, chrysoberyl, garnet, fluorite, lepidolite, topaz, tourmaline.

Extrusive — Volcanic

Rock Type	Landscape Form	Description	Collectable Minerals
Rhyolite	Flows.	Light to medium brown color. Composed of feldspar and silica. Uniform flowing texture.	Obsidian, opal, pitchstone, pumice, topaz, tourmaline.
Trachyte	Flows.	Light banded color. Feldspar. Flowing, "trachyte" texture.	Opal, turquoise.
Basalt	Flows, columns, cliffs, dikes.	Dark color. Composed of olivine, plagioclase, and pyroxene. Most common volcanic rock.	Calcite, chalcedony, hematite, jasper, opal, quartz, prehinite, topaz, tourmaline, zeolite.
Andesite	Flows, cliffs.	Light to dark brown color. Orthoclase and andesine feldspar.	Sulfur. Usually nothing else of interest for collectors.
Tuff	Hills, small mountains, outcrops.	Light tan color. Consolidated ash.	Chalcedony, geodes, fire-agate.

the ground, new secondary mineral crystals are formed such as wulfenite and cerussite in lead ore bodies, and azurite and chrysocolla in copper ore bodies. Whole suites of other minerals quietly crystallize out in underground cracks and crevices as if they were organic colonies of blossoms growing under the protection of the host rock. Over time, several successive crystallization episodes may occur, like crystal growing seasons, each cultivating its own species. The resulting crystal bouquets are very desirable and are avidly harvested by mineral collectors. The crystals produced by this calm subdued weathering process tend to be much smaller and more perfectly formed than the larger gnarlier crystals forged in the crucibles of igneous pegmatites. Chemical weathering is also responsible for the mineralized percolating water that fills hollow geodes and vugs with sharp bright crystals, turns volcanic bombs into agate filled thunder eggs, and supplies the calcium and silicon that turns mud balls into septarian nodules. Voids become crystallized inwardly from the void surfaces to the center leaving concentric layers of colorful mineralization and often culminating in a burst of sharp crystals in the center.

    The forces that turn mountains into beach sand can be much more forceful and dramatic than the gentle process described above. As mountains are eroded some of the rocks and minerals contained therein are durable enough to withstand the forces that tore them loose, tumbled then down mountain sides, and compacted them under thick layers of rubble. Dry washes and alluvial fans at the base of mountain ranges and the steep valleys that spill out of them may

TABLE 10
Sedimentary Rock Mineralogy

Rock Category	Rock Type	Morphology	Collectable Minerals
Terrigenous (Latin for earthborn)	Conglomerates, breccias, sandstones, mudstones, clay	Decomposition, disintegration, cementation	Weathered minerals, metallic ores, secondary minerals. calcite, chalcedony, hematite, muscovite, opal, rutile
Allochemical (biochemical/ biogenic)	Limestones, dolostones, phosphorites, chert, coal	Organic precipitation, movement, redisposition	Aragonite, beryl, calcite, dolomite, flint, fluorite, jet, opal, quartz
Orthochemical (inorganic chemical)	Evaporates, chert, travertine, iron formations	Chemical precipitation, evaporation of mineral water, percolation.	Borates, carbonates, chlorides. Alabaster, anhydride, aragonite, celestite, goethite, glauberite, halite, hematite, satin spar, travertine.

inerals such as agate, schorl, quartz crystal, [brok]en off and tumbled down from above. These [end up] on in sedimentary conglomerate rock strata. [...co]mbine with other minerals and precipitate [as l]ayers of travertine "onyx". Aragonite and [gen]eration in sedimentary limestone deposits. [...]ic minerals are popular lapidary materials. [...]y eventually grind up the softer minerals [becom]e suspended or dissolved in water. As the [...] the ground, these minerals are redeposited [...]alters magnesium silicates in the igneous [...m]ineral serpentine is deposited. When [...] the serpentine then becomes known as [...] for the mineralogy of sedimentary rock

***rphosis* for change form)**

[...]nentary, and to a lesser extent, igneous [...] sink so deeply into the earth's crust that [enough heat and] pressure is generated to alter their chemical composition, crystal structure, and physical appearance. It is like firing a pot in a kiln. You metamorphose clay into glass. Contact metamorphism occurs when the lowest stratum of the sedimentary column is pressed down so far that it comes in contact with the underlying magma causing it to intrude into the sedimentary rock. The reaction caused by the contact of magma and sedimentary rock results in extensive mineralization. These contact zones are fertile hunting grounds for mineral collectors. Erosion and geologic forces sometimes bring them to the surface. Keep an eye out for them. The minerals contained in metamorphic rock are determined by two factors. The first is the parent rock called the protolith. You cannot make a silk purse out of a sow's ear. Copper based minerals cannot be fathered by protoliths that contain no copper. Based on protolithic mineral content, there are four groups of metamorphic rocks. Pelitic rocks are parented by clay, mudstone, and sandstone sediments containing large amounts of aluminum. Calcareous rocks come from limestone sediments containing large amounts of calcium. The malfic group is from basalt and gabbro containing iron, manganese, and calcite-rich feldspar. The quartzo-feldspatic group is derived from quartz and feldspar and maintains primarily the same composition in the metamorphic form. The second factor that determines the mineralization of metamorphic rock is the amount of heat generated by the process. Depending on the temperature, pelitic rock may contain slate, schist, andalusite, kyanite, muscovite, staurolite, almandine garnet, and others. The calcareous group can contain grossularite garnet, wallastonite, actinolite, diopside, epidote, talc, and

TABLE 11
Metamorphic Rock Protoliths

Protolith	Metamorphic Rock Type
Shale	Slate, phylite, schist.
Basalt	Schist
Granite	Gneiss (pronounced "nice")
Limestone	Marble
Sandstone	Quartzite
Peridotite	Serpentine
Conglomerate	Metaconglomerate
Chert	Metachert
Coal	Graphite

others. Among the malfic group may be almandine garnet, hornblende, pyroxene, chlorite, diopside, epidote, and others. See Table 11 for protoliths of metamorphic rock types and Table 12 for a list of metamorphic minerals of possible interest to collectors.

Mineral Classification

James Dana's *System of Mineralogy* is the standard method of classifying minerals. It is based on the chemical composition of minerals, which are essentially elements or chemical compounds. Although there may be several varieties of a given mineral such as quartz and garnet, all varieties are chemically alike. The minerals in each group bear a strong chemical resemblance to one another and tend to be found together in the same geologic environment. According to the Dana system, the broadest category is the class, which is subdivided into families based on chemical comparison. Families are then subdivided into groups based

TABLE 12
Metamorphic Minerals of Interest to Collectors

Actinolite	Danburite	Orthoclase	Spinel
Almandine garnet	Diopside	Pyrite	Staurolite
Apatite	Dolomite	Rhodinite	Talc
Bixbyite	Epidote	Rhodochrosite	Tourmaline
Brookite	Kyanite	Rutile	Zircon
Calcite	Magnesite	Scheelite	
Chalcopyrite	Mica	Serpentine	
Chromite	Nephrite jade	Siderite	
Corundum	Olivine	Spessartine garnet	

on crystal structure. Groups, like plants and animals, are further subdivided by individual species that may have several varieties. Table 13 includes only select minerals that may be of interest to the average collector. The order of the chemical classes listed in the table is the same as the Dana system. However, since not all classes have been included, the consecutive number of each class does not necessarily correspond to the Dana listing. If there is no entry under the family column, then none was listed in Dana. Many groups carry the same name as one of its species. Some groups have only one species listed. The table below was compiled from a display of the Dana system presented at the University of Arizona Mineral Museum in Tucson, Arizona.

TABLE 13
The Dana System of Mineral Classification

Chemical Class	Family	Group Chemical Formula, Crystal System	Species
1. Native Elements	Metals	Gold, Au, isometric	Copper, gold, lead, silver
		Platinum, Pt, isometric	Iridium, osmium, palladium, platinum
		Iron-nickel, Fe, Ni, isometric	Chromium, iron, nickel, tin
	Semi-metals	Arsenic, As, hexagonal	Antimony, arsenic, bismuth
	Non-metals	Sulfur, S, orthorhombic	Diamond, graphite, silicon, sulfur
2. Sulfides	Sulforsenides	Chalcocite, Cu_2S, hexagonal	Chalcocite (copper sulfide)
	Arsenides		
	Tellurides		
		Galena, PbS, cubic	Galena
		Sphalerite, ZnS, isometric	Sphalerite
	Selenides	Chalcopyrite, Cu_2S_3, tetragonal	Chalcopyrite
		Stibnite, Sb_2S_3, orthorhombic	Stibnite
		Pyrrhotite, FeS, hexagonal	Pyrrhotite (magnetic pyrite)
		Cinnabar, HgS, hexagonal	Cinnabar

Chemical Class	Family	Group Chemical Formula, Crystal System	Species
2. Sulfides (cont.)			
		Realgar, As_4S_4, monoclinic	Realgar
		Orpiment, As_2S_3, monoclinic	Orpiment
		Pyrite, FeS_2, orthorhombic	Pyrite
		Marcasite, FeS_2, orthorhombic	Marcasite
3. Oxides	Simple oxides	Cuprite, Cu_2O, isometric	Cuprite
		Hematite, Fe_2O_3, hexagonal	Hematite, ilmenite
		Zincite, ZnO, hexagonal	Zincite
		Rutile, TiO_2, tetragonal	Cassiterite, pyrolusite, rutile, uraninite.
		Corundum, Al_2O_3, trigonal	Corundum
	Multiple oxides	Spinel, XY_2O_4, isometric.	Chromite, chrysoberyl, magnetite, spinel.
4. Hydorxides		Brucite, $Mg(OH)_2$, hexagonal	Brucite, magnatite
		Diaspore, AlO·OH, orhtorhombic	Geothite, diaspore
5. Halides	Anhydrous and hydrated	Halite, NaCl, isometric	Cryolite, fluorite, sylvite
	Compound	Creedite, monoclinic, $Ca_3Al_2So_4(OH_2)F8·2H_2O$	Creedite
6. Carbonates	Anhydrous	Calcite, $CaCO_3$, hexagonal	Calcite, magnesite, rhodochrosite
		Aragonite, $CaCO_3$, . orthorhombic	Aragonite, cerussite
		Dolomite, $CaMg(CO_2)_2$, trigonal	Ankerite, dolomite
7. Nitrates		Nitratite, $NaNO_3$, hexagonal	Niter, nitrate
8. Borates		Borax, monoclinic. $Na_2B_4O_5(OH)4·8H_2O$,	Borax, colemanite, ulexite

Chemical Class	Family	Group Chemical Formula, Crystal System	Species
9. Sulfates	Anhydrous and acid	Barite, $BaSO_4$, tetrahedral	Anglesite, anhydrite, celestite, crocoite
		Glauberite, $Na_2Ca(SO_4)_2$, monoclinic	Glauberite
	Hydros	Gypsum, $CaSO_4$, monoclinic	Selenite, satin spar, gypsum
	Anhydrous with hydroxyl or halogen	Brochantite, $CU_4(SO_4)(OH)_6$, monoclinic	Brochantite, linarite
10. Molybates and Tungstates	Anhydrous	Scheelite, $CaWO_4$, tetragonal	Scheelite, wulfenite
11. Phosphates, Arsenates and Vanadates	Anhydrous with hydroxyl or halogen	Apatite, $Ca_5(PO_4)_3(F,CL,OH)$, hexagonal	Apatite, fluorapatite, mimetite, pyromorphite, vanadinite
12. Silicates	Nesosilicates	Phenacite, Be_2SiO_4, hexagonal	Phenacite, willemite
		Olivine, $(Mg,Fe)_2SiO_4$, orthorhombic	Olivine, peridot
		Garnet, $Fe_3Al_2(SiO_4)_3$, isometric	Almandine, andradite, grossular, hydrogrossular, pyrope, spessartine, uvarovite, zircon
		Andalusite, Al_2SiO_5, orthorhombic	Andalusite, kyanite, sillimantite, staurolite, topaz
		Humite, $Mg_7(SiO_4)_3(F,OH)_2$, monoclinic	Chondrodite, datolite, humite
	Sorosilicates	Hemimorphite, $Zn_4(Si_2O_7)(OH)_2 \cdot H_2O$, orthorhombic	Hemimorphite
		Epidote, $Ca_2(Al,Fe)AlO_2(SiO_4)(SiO_7)(OH)$, monoclinic.	Allanite, epidote, Vesuvianite.
	Cyclosilicates	Beryl, $Be_3Al_2(Si_6O1_8)$, triclinic	Axinite, beryl, cordierite, tourmaline.
	Inosilicates	Pyroxene, XYZ_2O_6, orthorhombic	Diopside, enstatite, jadeite, spodumene

Chemical Class	Family	Group Chemical Formula, Crystal System	Species
12. Silicates (cont.)			
		Pyroxenoid group, triclinic	Rhodonite, wallastonite.
	Amphebole,	Actinolite, anthophy-lite, hornblende Wo$_{-1}$X$_2$Y$_5$Z$_8$O$_{22}$(OH,F)$_2$, orthorhombic	
	Phyllosilicates	Serpentine, MgSi$_2$O$_5$(OH)$_4$, monoclinic	Antigorite, chrysotile
		Clay	Kaolinite, talc, pyrophyllite
		Mica, KAl$_2$(AlSi$_3$O$_{10}$)(OH)$_2$, monoclinic	Biotite, lepidolite, muscovite
		Chlorite, (Mg,Fe)$_3$(Si,Al)$_4$(OH)$_2$·(Mg,FE)$_3$(OH)$_2$, monoclinic	Apophyllite, chlorite, chrysocolla, prehinite
	Tectosilicates	SiO$_2$ Group, Hexagonal	Opal, quartz, tridymite
		Feldspar, KaiSi$_3$O$_8$, monoclinic and triclinic	Albite, danburite, microcline, orthoclase
		Feldspathoid, KAiSi$_3$O$_8$, trigonal	Lazurite, leucite, sodalite
		Scapolite, (Cl$_2$,SO$_4$,CO$_3$)$_2$(Al$_2$Si$_2$O$_8$)$_6$, trigonal	Analcine, marialite, meionite
		Zeolite, monoclinic	Chabazite, heulandite, natrolite, stilbite
	Quartz	Quartz, SiO$_2$, trigonal	Chalcedony, aventurine, amethyst, chrysoprase, citrine, carnelian, flint, jasper

The Mineral Collecting Sites

SITE 1

Selenite Near Dutchman Wash

Difficulty Scale: 2 – 3 – 1   Seasons: All

Global Positioning System Coordinates: 36° 58' 42.3" N, 113° 27' 50.2" W*

Geology: Middle-Early Triassic Sedimentary Moenkopi Formation

U.S. Geological Survey 7.5 Minute Topographical Map: Yellowhorse Flat

ALTHOUGH THIS SITE IS LOCATED IN ARIZONA, it is best approached from the town of Washington, Utah, located on Route 9 between Saint George and Hurricane. From Interstate 15, take exit 10 to Washington. Turn right (south-east) at the end of the off ramp and go .1 mile to the traffic light and turn left (north-east) onto Telegraph Road. Drive 1 mile on Telegraph Road and turn right (south-east) on 300 East which becomes Washington Fields Road. Proceed 3.7 miles to a T and turn left (east) onto 3650 South. Go .2 mile and turn right (south- east) on KD-JO Lane. The distance to the site from here is 8.2 miles. After going 2.5 miles, the pavement ends at a cattle guard and at the 6.3 mile point, you will come to a fork in the road at Fort Pearce Wash. Take the right fork across the wash. The road then becomes BLM Road 1035. Continue on for another 1.9 miles to the collecting site. The roads are well maintained and suitable for passenger cars. When you have driven 8.4 miles from the T, you will see a little selenite quarry on the right hand side of the road.

Collecting here is a picnic. The quarry is about the size of a baseball diamond and about 30 feet deep at its lowest point. At the back side of the quarry is a solid vein of clear glassy selenite approximately 15 – 20 feet high. On a sunny day, it looks like a glacier. The quarry is chock full of piles of selenite windows 4 – 12 inches square and 1 to 3 inches thick. The faces are smooth and clear as glass although, the edges may be dirt encrusted. They are so clear that you can read the page of a book through them. You can sit down anywhere and hi-grade as long as you like. If you are a little more energetic, and desirous of larger and chunkier specimens, you can try to extract larger pieces out of the vein in the back wall. Although selenite specimens are rather fragile and scratch easily, this large solid vein is surprisingly tenacious. Breaking pieces off is tough work, but once you get your chisel moving in the right direction along a cleavage plain, you can slough off larger pieces fairly easily. A lot of hammer and chisel work, however, will yield little more than a pile of white gypsum dust. Look for vugs and crystally horizontal openings. Try to remove pieces above and below the openings. If successful, your efforts will yield a very attractive, clear, chunky piece with one snowy, crystally face.

There are several more selenite quarries between Hurricane Cliffs to the East, Black Rock Canyon to the west, and Wolf Hole to the south. Some of these seem to be abandoned and others appear to be in operation from time to time. Be sure to honor any claim markers or no trespassing signs you may encounter. There is a huge amount of territory to explore here in the Arizona Strip between the Utah border and the Grand Canyon. Before doing any collecting, however, be sure you know where you are. Much of this area is contained within several wilderness areas, recreational areas, and national parks.

The selenite quarry by Dutchman Wash.

*G.P.S. coordinates taken at the quarry.

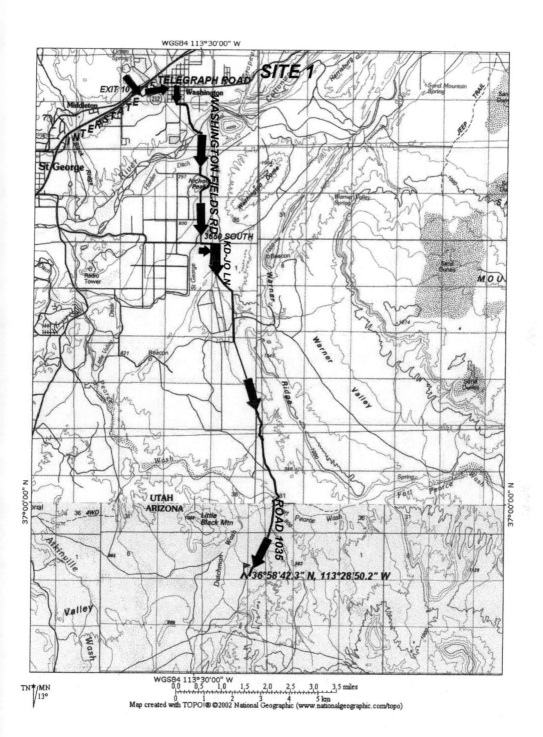

SITE 4

Hematite at the BBC Mine

Difficulty Scale: 6 – 5 – 6   Seasons: Fall, Winter, Spring

Global Positioning System Coordinates: 34° 10' 00" N, 113° 56' 04" W*

Geology: Early Miocene-Oligocene Metamorphosed Schist and Gneiss

U.S. Geological Survey 7.5 Minute Topographical Map: Planet

View of the BBC Mine from the tailings pile on the asphalt pad.

FROM BOUSE ON STATE ROUTE 72, follow the directions to Midway on page 60. From Midway, go north on Swansea Road 5.5 miles to a 4-way stop. Continue straight (north) through the intersection onto Planet Ranch Road, drive 1.4 miles, and turn right (north-east). From here, you can see the BBC Mine excavations on the large hillside directly ahead of you. Follow the road .7 mile midway up the hill where you can park and turn around. Three short roads cut horizontally across the hillside forming three levels. Walk a few hundred feet to the collecting area on the second level. On the way to the hill, you will pass by a tailings dump on an asphalt pad. You can also collect hematite here if you wish.

Hematite is a common mineral throughout Arizona and is particularly prevalent here in the Buckskin Mountains. Generally speaking, it is not a highly sought after mineral by collectors. It is black and usually dirty. However, the hematite at the BBC mine and its environs is gem quality. You can collect pieces of massive hematite from one to six inches across that prominently display clusters and arrays of brilliant, bladed, jet-black, sharp-edged crystals. The faces of these crystals have a mirror finish that make them look as if they were manufactured and polished in a machine shop. As an added bonus, the spaces between the crystals are often filled by clear acicular quartz crystals. And, some specimens are also accented with zones of forest green colored lepidolite and adorned with sky blue chrysocolla lenses. The combination of all the minerals create some very attractive and interesting cabinet specimens.

You can collect in the dumps on the slope between levels one and two or you can work in the banking on the uphill side of level two. In the dumps, you can sort through the rubble in search of collectable pieces. Or, you can break up and trim larger pieces to create suitable specimens. At the slope on level two, carefully dig into the dirt to extract pieces of hematite for further examination.

G.P.S. coordinates taken at the second level.

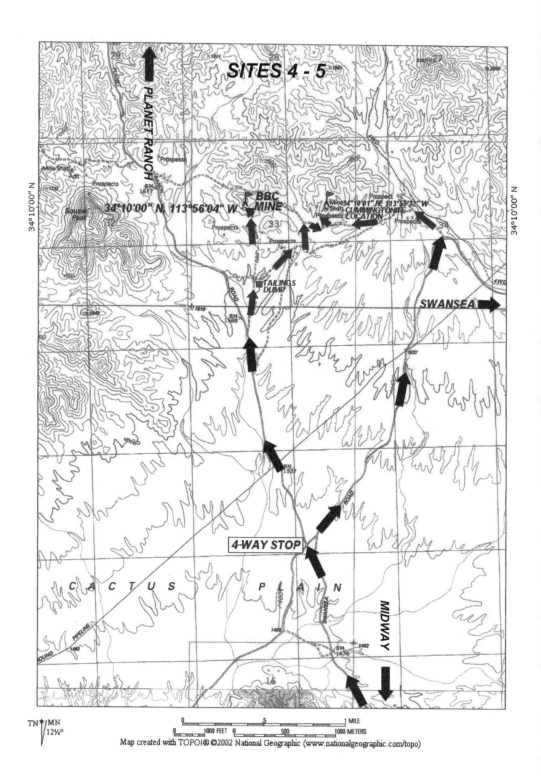

SITE 5

Cummingtonite East of the BBC Mine

Difficulty Scale: 6 – 5 – 6   Seasons: Fall, Winter, Spring

Global Positioning System Coordinates: 34° 10' 01" N, 113° 55' 32" W*

Geology: Early Miocene-Oligocene Metamorphosed Schist and Gneiss

U.S. Geological Survey 7.5 Minute Topographical Map: Planet

Approaching the collecting area from the east.

FOLLOW THE DIRECTIONS to the BBC Mine on page 50. As the crow flies, the mine shaft and tailings pile containing the cummingtonite is .6 mile directly east of the BBC Mine. Even though you can easily see the location from the BBC Mine, the route over to it is somewhat difficult to find because there are so many faint old roads and paths running in all directions through the washes and over the hills to several prospects scattered throughout the area. There are more roads on the ground than there are on the map. There are two approaches. First, when you leave the BBC Mine, you can go back to the tailings dump and follow the faint road eastward that intersects with other roads that lead to the collecting area. Or, you can return to the 4-way stop, turn east onto Swansea Road, go 2.1 miles, and turn left (north). Follow this road north .3 mile, turn left (west), and go .2 mile and turn right (north-west). At this point, you can see the tailings on the hillside across the ravine. Follow this road .5 mile down into the wash and up to the mine shaft where you can park. Collect in the hematite tailings surrounding the mouth of the mine shaft.

This site is only for those who like to collect obscure and esoteric minerals. Cummingtonite is a monoclinic amphibole that usually forms in metamorphic and sometimes in igneous rock. Cummingtonite forms in a capillary habit that gives it an organic rather than a mineral appearance. Look for a tangle of golden-brown hair-like masses growing in the cracks and crevices of the hematite rubble. The geologic formation at this site is Pinal Schist and Yavapai Supergroup. The cummingtonite forms in hematite matrix. Since the hematite here is fairly solid with small vugs and narrow seams, cummingtonite occurrence is sparse and formations are small. The easiest collecting method is to carefully examine hematite pieces in the rubble pile with a loop or magnifying glass for cummingtonite. Or, you can break open pieces with your rock pick in search of cummingtonite bearing vugs. Do not expect to discover more than one or two decent small specimens. There are several more prospects and hematite outcrops scattered throughout the hillsides and washes within a mile or so of the site. You may find more cummingtonite occurrences in some of these locations.

G.P.S. coordinates taken at the collecting site.

SITE 6

Chrysocolla along Rankin–Lincoln Ranch Road

Difficulty Scale: 7 – 6 – 6   Seasons: Fall, Winter, Spring
Global positioning Survey Coordinates: 34° 03' 53" N, 113° 47' 17.4" W*
Geology: Early Miocene-Oligocene Granitic and Granodiorite Rock
U.S. Geological Survey 7.5 Minute Topographical Map: Swansea

The collecting area viewed from the cabin.

FROM THE TURNOFF TO THE GOLD HILL MINE (Site 7, page 58), continue northeastward on Rankin–Lincoln Ranch Road 2.35 miles and turn left. This turn is another faint hard to spot road. You will need 4-wheel drive immediately to get up the steep, eroded incline that wraps up and around the hillside. As you curve up and around to the west, you will come to a fork. One road goes up to the top of the ridge. The other goes down into a little valley where you can see the ruins of a small, wooden cabin. Drive down to the cabin and park. The distance from Rankin–Lincoln Ranch Road to the cabin is only .25 miles. Getting to the cabin is the easy part. The rugged peak on the north side of the cabin is the collecting area. Before you hike up the steep, slippery slope, walk up the road beyond the cabin a few hundred feet until the excavations on the west side of the peak come into view. This will give you a good idea of what awaits you at the collecting site.

Outside of the huge, open-pit, commercial copper mines, this may be the best chrysocolla collecting location in Arizona. Viewed from below, the tailings and excavations at the top of the peak look as if they had been painted blue. There is an abundance of material here. Above and around the mine adit entrances, rich deep blue chrysocolla seams course through the rock faces. The ground is littered with small blue rocks and chips. The best material is in the tailings. Rocks up to a foot across contain chrysocolla veins ¼ – 1 inch wide. Some of these veins are solid, pure, and of gem quality. Others are vuggy, bubbly seams containing calcite as well as chrysocolla.

The climb up to the collecting area from the cabin is not a great distance, but it is very steep and strenuous. Keep in mind that whatever you collect, you will have to lug down on your back. So, bring your hammer and chisel to field trim as much matrix off your specimens as you can in order to lighten your load as much as possible.

*G.P.S. coordinates are taken at the cabin.

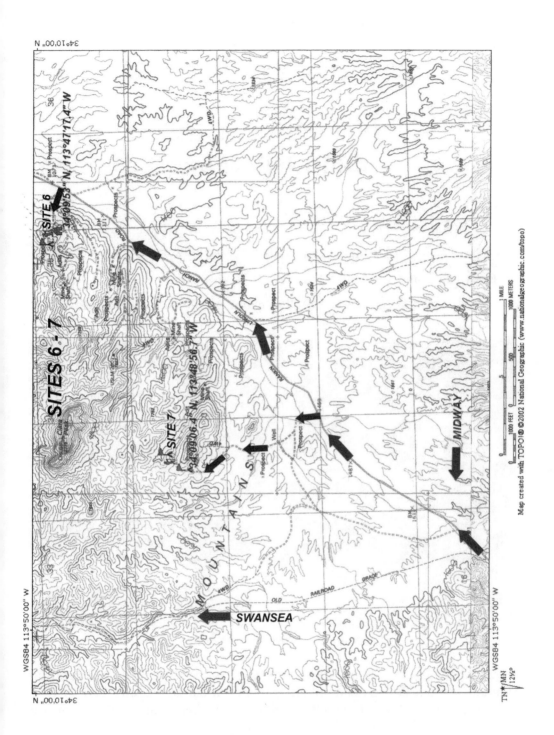

SITE 7

Malachite at the Gold Hill Mine

Difficulty Scale: 7 – 5 – 6   Seasons: Fall, Winter, Spring

Global Positioning System Coordinates: 34° 09' 06.4" N, 113° 48' 56.7" W*

Early Miocene-Oligocene Metamorphic Granitic and Granodiorite Rock

U.S. Geological Survey 7.5 Minute Topographical Map: Swansea

The road to the Gold Hill Mine.

FOLLOW THE DIRECTIONS FROM BOUSE TO MIDWAY on page 60. From Midway, take Rankin–Lincoln Ranch Road 6.7 miles to an un-maintained dirt road on your left leading north toward Clara Peak. About a mile before the turn, you can see mine shafts and tailings slides on your left (north-east) high up on the steep side of the ridge in front of Clara Peak. Your destination is the lowest mine on the west end of this ridge. Look carefully for the turnoff. The road is faint and obscured by the berm left by the road grader along the edge of the road. You can drive all the way up to the mine. The distance is 1 mile. Although, the last .2 mile is an easy 4-wheel drive. You can park and turn around at the first excavation. The road continues upward and around the back side of the ridge. Do not drive up there, however, because there is no room to turn around.

This is an excellent malachite specimen collecting site. In fact, it may very well be the best of its kind in Arizona. You can collect solid 1/4 – 1/2 inch gem quality malachite veins, small vugs filled with fibrous malachite, and botryoidal malachite lined seams accompanied by calcite and barite crystals. Look for malachite bearing rock up to a foot across in the tailings. Trim the rocks carefully with your rock pick to expose the vugs and seams within. Since vugs and seams are the weakest part of a rock, it will tend to crack or split along the seam or through the vug when struck. So, in order to retrieve the best specimens, be patient and wield your hammer gently.

Up the road another 300 – 400 hundred yards on the north-east side of the ridge is another collecting location. It is an easy walk up the road from the parking area. Here, you will find an excavation yielding a good amount of light-blue chrysocolla. Bring a hammer and chisel along with your rock pick so you can trim out the best specimens.

G.P.S. coordinates taken at the malachite location.

SITE 8

Augite at the Green Streak Mine

Difficulty Scale: 5 – 5 – 3   Seasons: Fall, Winter, Spring

Global Positioning System Coordinates: 34° 06' 00" N, 113° 55' 41" W*

Early Miocene-Oligocene Metamorphic Granitic and Granodiorite Rock

U.S. Geological Survey 7.5 Minute Topographical Map: Powerline Well

View of the road beside the dump leading up to the collecting area.

FROM THE INTERSECTION OF Main Street and State Route 72 in Bouse, turn east on Main St., cross the railroad tracks, and drive .25 mile to Rayder Road. Turn left (north) on Rayder, which soon becomes Swansea Road, and go 2.2 miles to a fork. Bear left (north-east) and drive 10.5 miles to Midway, which is the intersection of Swansea, Rankin–Lincoln Ranch, and Transmission Line Roads. Follow Swansea Road northward 2.2 miles and turn left (west) onto an unmaintained dirt road. Follow this road for 2 miles. You will see the tailings pile

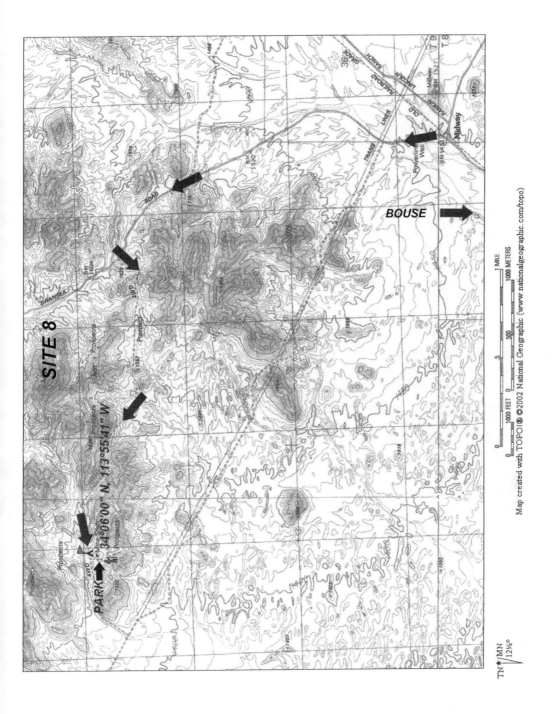

beside the road on your left (south) shortly after passing the remains of a small, old, stone house and crossing a wash. There is a road to the left (south) on the west side of the pile where you can turn in and park.

There are some average quality cooper minerals in the mine dumps here. You can see malachite, chrysocolla, and perhaps a little brochantite attached to the small rocks in the tailings pile. The most interesting material at this site, however, is on the road leading up to the prospect on top of the hill directly south of the mine. This is the same road where you turned in and parked. Walk a few hundred feet farther down this road (south) to the wash. Here, you will begin to see pieces of augite scattered on the ground. As the road ascends vertically up the hill, it cuts across a large horizontal vein of augite. Augite is a fairly common rock forming mineral present in many igneous formations such as gabbro, diabase, and basalt. It is usually of little interest to the collector or lapidary unless it has some esthetic quality. Material that was plowed up by the bulldozer is available in the road bed and in the roadside berms. Complimenting the augite are veins of common light green epidote that is often found throughout southwest Arizona. Some of this is adorned with small crystals. The real attraction, however, is the dark, rich, gemmy, massive augite. There are large chunks of this material up to a foot across. Some pieces have wide white quartz and common light green epidote veins running through them. This material makes an excellent lapidary stone. It cuts and shapes well and takes a brilliant polish.

*G.P.S. coordinates taken at the mine.

SITE 9

Minerals South of Plomosa Road

Difficulty Scale: 3 – 4 – 4   Seasons: Fall, Winter, Spring
Global Positioning System Coordinates: Given with each location
Geology: Jurassic-Cambrian Metamorphosed Sedimentary Rock
U.S. Geological Survey 7.5 Minute Topographical Map: Ibex Peak

Location #1, Phoenix and Yuma Mine,
G.P.S. coordinates: 33° 48' 25" N, 114° 03' 59" W*

Looking up from the wash at the tailings
pile at the top of the ridge at location #1.

FROM THE INTERSECTION OF MAIN STREET and State Route 95 in Quartzsite, go north on S.R. 95 5.7 miles to Plomosa Road, also known as the Bouse-Quartzsite Road. Turn right (north-east) and drive 12.7 miles to an unmarked gravel road on your right. If you are coming from Bouse, turn east on Joshua Road which intersects Plomosa Road. Total distance from Joshua Road to the gravel road is 9.4 miles. Turn south, go .1 mile, turn south again, and drive .5 mile to a fork. Take the left fork that goes down into the wash. As you enter the wash, you will see the main tailings pile on the ridge to your right. Park in the wash and hike up the washed out road to the top of the ridge. On the way up, you will pass an excavation on your right containing an exposed vein of solid granular hematite. When you reach the crest of the ridge, the main tailings pile and vertical shaft will be on your left. There is nothing much to collect there. Directly in front of you, the west side of the ridge slopes down into a wash. This slope is littered with rocks containing hematite, quartz, epidote, calcite, and jarosite. In the west banking across the wash is a small dig containing gypsum. On the hill above the gypsum deposit is another prospect containing hematite in conjunction with malachite and chrysocolla. Most of this material can be easily picked up off the ground. You may want to bust up some of the larger rocks with your rock pick to reveal even better samples.

The minerals in this area are interesting and colorful. The gypsum comes in two forms. One comes in layers that resemble hard packed, pure, clean, white snow both in color and texture. The other form looks like delicate three-dimensional frozen lace. The hematite varies in texture and consistency. One type is massive and granular. When you touch it, it will leave a smear of hematite

View of the Mudersbach Mine from the road.

dust on your hand. It is very messy and dirty. Another habit is in the form of small bladed druse. This type is clean, hard, and has a very brilliant luster. A third type is larger blades and chunks that have a very shiny metallic luster. Mounted on colorful matrixes, combinations of these minerals make excellent specimens. Pieces displaying chunky hematite, crystally epidote, milky-white calcite, and golden jarosite druse are the most desirable. Samples showing streaks of green malachite and blue chrysocolla adorned with shiny hematite druse are also very striking.

*G.P.S. coordinates are at the parking area in the wash.

Location # 2, Mudersbach Mine,
G.P.S. coordinates: 33° 48' 13" N, 114° 03' 38" W**

To reach location #2, follow the road .5 mile from the parking place in the wash at location #1 to the Mudersbach Mine. The narrow road dips in and out of washes as it curves its way eastward to location #2. The collectable material here is similar to that found at location #1 except that it contains more malachite and chrysocolla.

** G.P.S. coordinates are taken at the Mudersbach Mine.

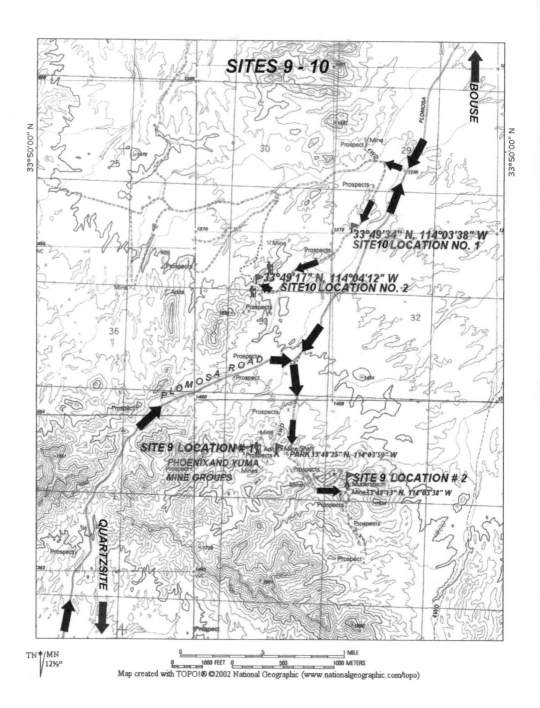

SITE 10

Barite West of Plomosa Road

Difficulty Scale: 4 – 4 – 6   Seasons: Fall, Winter, Spring

Global Positioning System Coordinates: 33° 49' 17.3" N, 114° 04' 12" W*

Geology: Jurassic-Cambrian Metamorphosed Sedimentary Rock

U.S. Geological Survey 7.5 Minute Topographical Map: Ibex Peak

FROM THE INTERSECTION OF Main Street and State Route 95 in Quartzsite, go north on S.R. 95 5.7 miles to Plomosa Road, also known as the Bouse-Quartzsite Road. Turn right (north-east) on Plomosa Road, drive 12.2 miles, and turn left (west) on the unmarked gravel road. En route, you will pass the turnoff to the Phoenix-Yuma and Mudersbach Mines (see site no. 9 on page 63). After turning left, travel .1 mile on the unmarked gravel road and turn left (south-west) again. At the .6 mile point, you will come to location #1 (33° 49' 34" N, 114° 03' 48" W), a trench mine on your right that parallels the west side of the road. This excavation follows a seam of barite and other minerals similar in appearance to the material at location #2 farther down the road. From location #1, proceed down the road another .6 mile to a rough, narrow, little road on your right (west). Follow this road up to the collecting site at the top of the ridge. There is room there to park and turn around.

This site contains perhaps the most exquisite lapidary material in Arizona. Along the top of the ridge is a large, bright, gemmy, colorful vein running through the dull, earthy, igneous country rock. It looks like a giant piece of ribbon candy

Location #1. The trench mine beside the road en route to location #2.

following the undulations of the ridge line from north to south. Minerals represented in this vein include transparent quartz, black hematite, white barite, red jasper, and orange and yellow agate. These minerals are arranged in thin wiggly parallel bands. The barite and jasper layers tend to be the widest. There are also veins of white, massive, crystalline barite several inches thick accompanying and intersecting the main vein. Drusey vugs and open seams are also present.

The width of the vein varies from several inches to several feet. The slopes on both sides of the ridge are littered with pieces that have eroded or been broken out of the deposit. So, there is plenty of loose material to choose from. Or, you can attempt to liberate material from the main vein with hammers, chisels, or crow bars.

The road leading up the ridge to the main vein.

G.P.S. coordinates taken on the road at the top of the ridge.

SITE 11

Epidote Near Boyer Gap

Difficulty Scale: 6 – 5 – 3   Seasons: Fall, Winter, Spring

Global Positioning System Coordinates: Given with each location

Geology: Jurassic-Cambrian Metamorphosed Marble and Quartzite

U.S. Geological Survey 7.5 Minute Topographical Maps:
Quartzsite and Middle Camp Mtn

FROM THE INTERSECTION OF Main Street and State Route 95, (Central Boulevard), in Quartzsite go north 1.6 miles and turn left (west) on Tyson Street. Go .7 mile on Tyson St. and turn right (north) on North Desert Avenue. After 1 block, the pavement ends and North Desert Ave. becomes Moon Mountain Road although, there is no sign announcing the name change. Follow Moon Mountain Road to location #1. The distance is 7.3 miles from the intersection of Tyson St. and North Desert Road. You can easily make it to location #1 in a passenger car.

Location #1, Boyer Gap, G.P.S. coordinates:
33° 44' 03.8" N, 114° 18' 56.5" W

View of the location #1 excavation from the road looking east.

Location #1 is the first low road cut you come to on the north side of the newly widened and graded Moon Mountain Road as it begins its ascent up Boyer Gap. Park off the road on the flat area immediately before the road cut so as to avoid the rock trucks from the Sunset Marble Quarry farther up Boyer Gap. The collecting area is the low road cut cliff and the wide slope behind it. In the face of the road cut cliff, you will see excavations where collectors have removed material from quartz-epidote veins that course through the earthy metamorphic rock. You can continue to harvest specimens from these veins or search the slope above for new outcrops. Look for shiny, long, dark green, columnar epidote crystals encased in white quartz. The rugged Dome Rock Mountains north of Boyer Gap are laced with strata containing quartz, epidote, marble, hematite, and copper minerals. The canyons and washes leading into these mountains, however, are steep and narrow, registering an 8 or a 9 on the difficulty scale.

Location #2, Dome Basin Mine, G.P.S. coordinates: 33° 44' 51" N, 114° 20' 22" W

View of the Dome Basin Mine from the intersection of roads 264 and 361.

To reach location #2, continue west on Moon Mountain Road from location #1. Within .2 mile, you will come to the gated entrance to the Sunset Marble Quarry. The road appears to end here at a no trespassing sign. Actually, the road maintenance ends at this point. The old road jogs to the left (south) around the corner of the gate and continues westward over Boyer Gap parallel to and outside of the chain-link fence that marks the southern border of the quarry property. At this point, however, the road becomes a 6 on the difficulty scale requiring

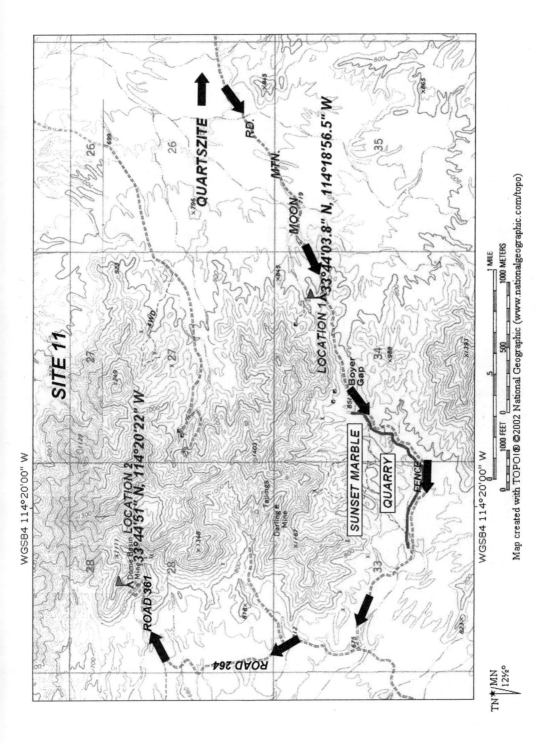

high clearance. From here, it is a slow tedious journey to the Dome Basin Mine. Go 2 miles west from the quarry gate and turn right (north) at the Road 264 sign. Follow this road 1.1 miles to road sign 361. At this point, you will see the Dome Basin Mine ruins on the mountainside to your right (east). Turn right and drive .4 mile to the bottom of the tailings slide.

Search through the tailings piles for specimens displaying multiple minerals. As at location #1, there are pieces containing quartz with epidote zones. This location also contains chunks of massive, dark green epidote coated with lighter and brighter green epidote druse. Some of this massive epidote is laced with shiny, black, metallic looking hematite. Additionally, since this was a copper mine, you may find some of the above specimens adorned with layers of chrysocolla and malachite. You may want to break open some of the more promising looking larger rocks and boulders with your rock pick or sledge to reveal the fresh, bright and colorful mineralization within.

SITE 12

Alunite on Sugarloaf Peak

Difficulty Scale: 5 – 4 – 6   Seasons: Fall, Winter, Spring

Global Positioning System Coordinates: 33° 38' 22.5" N, 114° 18' 49.8" W*

Geology: Jurassic Sedimentary and Volcanic Quartz-Feldspar Porphyry

U.S. Geological Survey 7.5 Minute Topographical Map:
Middle Camp Mountain

The intersection of the pipeline road and the road leading up
Sugarloaf Peak to the collecting area.

FROM INTERSTATE 10 WEST OF QUARTZSITE, take exit 11 onto Dome Rock Road and turn south-east. You will be going back toward Quartzsite on the south side of the freeway. After going .25 mile, turn right (south) onto Road 325. Go .1 mile to a fork with a 14 day camping limit sign and bear left onto the pipeline road. This is the same way you go to Site 13, page 76. Proceed east on the pipeline road for 1.25 miles where you will come upon a road to your right (south) leading up to the digs on Sugarloaf Peak which is sometimes mistakenly referred to as Dome Rock. You can drive .4 mile up this road and park at the switchback where you can turn around. Do not try to drive up the mountain any farther. From here, you can hike up to the highest excavation or across to the ones below it.

In the wall of the uppermost excavation, you can see veins of alunite which are often contained within larger veins of a similar looking mineral called dactite. These are sometimes separated from the country rock by thin, crusty sheets of selenite. These veins range in width from a half an inch to a foot. Look for veins containing red, parallel, fibrous looking streaks. Alunite is creamy-white, smooth to the touch, and opaque. It has a hardness of 3.5 – 4 making it a suitable material for lapidary work and statuary carving. It is solid, dense and has a uniform texture. When polished, it has a luster like porcelain or ivory.

If you are not interested in digging alunite out of the excavation walls, search through the rubble piles in the digs below the top level. Look for the pathways that descend to the lower levels. It appears that these excavations have only been highgraded leaving a fair amount of good alunite behind in the tailings.

*G.P.S. coordinates at the intersection of pipeline road and the turnoff to Sugarloaf Peak.

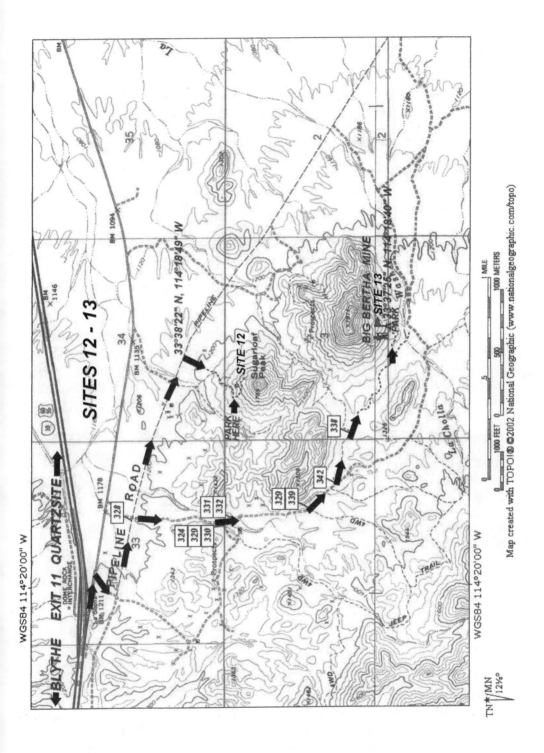

SITE 13

Bladed Hematite at the Big Bertha Extension Mine

Difficulty Scale: 4 – 4 – 6   Seasons: Fall, Winter, Spring

Global Positioning System Coordinates: 33° 37' 25.0" N, 114° 18' 40.0" W*

Geology: Jurassic Igneous Massive Quartz-Feldspar Porphyry

U.S. Geological Survey 7.5 Minute Topographical Map: Cunningham Mtn

THE DIRECTIONS TO THIS SITE ARE SOMEWHAT COMPLEX because the road grid in this area looks like it was designed by an arachnid. In fact, in the winter months, the place looks like a spider web adorned with snowbirds paralyzed and safely entombed in sheet metal cocoons. But if you follow the directions carefully, you should arrive at the collecting location without too much trouble.

The view of the mine coming down Road 339.

From Interstate 10 west of Quartzsite, take exit 11, onto Dome Rock Road and turn south-east. You will be going back toward Quartzsite on the south side of the freeway. After going .25 mile, turn right (south) onto Road 325. Go .1 mile to a fork with a 14 day camping limit sign and bear left (east) onto the pipeline road. Drive east .4 mile and turn right (south) on an unmarked road. Opposite the unmarked road, on the north side of the pipeline road, you will see road marker 328. Drive southward on the unmarked road .5 mile to the intersection of Roads 330, 324, and 329. Continue south on Road 329 down into a wash where you will encounter road signs 331 and 332 to the east. Ignore these and continue southward directly across the wash .4 mile to the intersection of Roads 329 and 339. Go south on Road 339 .3 mile and turn left (south east) onto Road 342. At this point, you can see the mine excavations on the mountain side to the south. Go .15 mile on Road 342 and turn right(south) on Road 338. After going .5 mile, turn left (east) on the faint, unmarked road and drive about .23 mile up to the tailings plateau where you can park. From here on, the road going up the mountain is steep, narrow, and rough. The excavation at the top is a very dangerous, severely undercut, fractured overhang. No need to go there. Even though it contains a vein of collectable material, it is just too dangerous to mine.

Instead of going farther up the mountain, hike across and up to the bottom of the tailings slide below the excavation. Look for solid, shiny, pure, snow-white quartz boulders containing brilliant, mirror-finish, jet-black hematite blades. This combination of gemmy black and white minerals make striking specimens. Bust up the large quartz rock with a heavy hammer to reveal the hematite. The quartz will trim up fairly well with your rock pick if you are careful and patient. More material is available on the bankings below the parking area and in the excavations to the east beside the wash. In the wash there are some huge boulders displaying excellent hematite on the surface. You would need a heavy sledge to harvest specimens from them.

*G.P.S. coordinates taken at the tailings plateau.

SITE 14

Geodes East of the Ramsey Mine

Difficulty Scale: 5 – 5 – 6   Seasons: Fall, Winter, Spring

Global Positioning System Coordinates: 33° 37' 34.9" N, 113° 58' 32.0" W*

Geology: Cretaceous-Late Jurassic Sedimentary Rock and Volcanic Rhyolite

U.S. Geological Survey 7.5 Minute Topographical Map: Bear Hills

FROM INTERSTATE 10, take exit 31, drive 3.8 miles east on U.S.60 toward Brenda, and turn right (south) on Ramsey Mine Road. Look for the road sign and a gate. Drive 3 miles on Ramsey Mine Road across I-10 to a fork. Bear left (east) and go .9 mile to a rough, narrow road leading up a hill to your right (west). Drive up the road .2 mile to the collecting area. This site shares a common boundary with the Ramsey Mine property on the west and north sides. Note the no trespassing signs on this patented claim.

The rough, narrow road leading up the hill to the collecting area.

This site has been well known for its geodes for many years and rightly so. However, it contains several other attractions as well. But first the geodes. Buried within a layer of clay and caliche is a bed of red and green rhyolite which encapsulates geodes and vugs one half an inch to two inches in diameter. In the past, a few geodes as much as 6 inches across have been reported. Some have weathered out and are loose in the soil. As with most geode beds of this type,

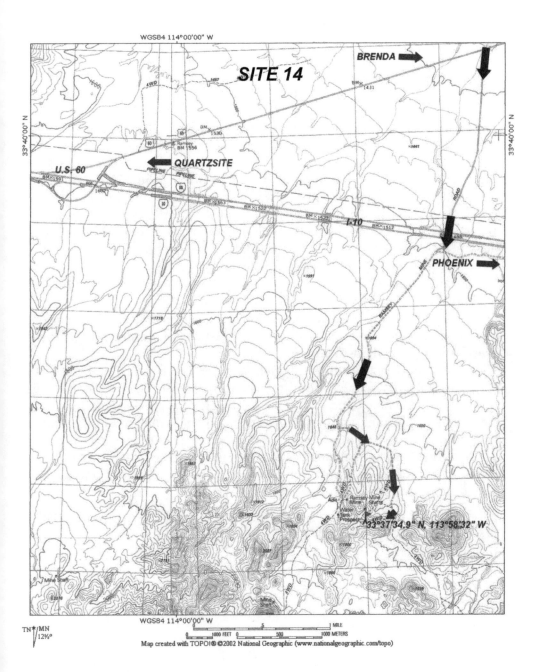

some are solid rhyolite, some are filled with clear quartz, and a few are hollow and crystally. The most highly prized are those containing sharp, clear, light amethyst crystals. Others contain gemmy, clear quartz crystal sprays and druses.

To collect these geodes, you must dig down 12 – 18 inches in the dirty, dusty, sometimes muddy clay. Once you extract the rocks at this depth you find that they all look alike because they are encrusted with caliche. It is literally a hit and miss affair. Even little round rocks that look for all the world like geodes often turn out to be merely little round rocks. Even though the rocks here including the geodes are hard and resilient, try striking them only as hard as necessary to crack then in two pieces. If you smack too hard you run the risk of destroying an excellent specimen. If you do not wish to dig, search through the tailings left by those who dug before. There is a lot of material exposed here. Not all of it has been thoroughly examined. By carefully searching through the debris, you may well discover some worthwhile pieces.

As a bonus, in addition to the geodes, you can also collect calcite and obsidian. The calcite is shiny, chunky, and jet black. Sometimes it comes in conjunction with the red and green rhyolite making a very attractive display. In the large pit dug into the hillside at the south-east end of the site is a thick vein of dark amber-colored obsidian.

The most unexpected find at this site, however, is the variety and quality of the fluorescents. The chalcedony lining the interior of the geodes and vugs glows a bright green. Some cavities contain cloudy calcite that fluoresces different shades of pink and red. Others contain agate that gives off a darker forest green. When they appear together, these minerals present a very pleasing array of color. The black calcite does not fluoresce, but the matrix it is attached to does. The country rock in the north-east corner of the site fluoresces an unusual rich maroon color and shades of pinkish-red.

*G.P.S. coordinates taken at the site.

SITE 15

Chalcedony on Hull Road

Difficulty Scale: 2 – 2 – 1   Seasons: Fall, Winter, Spring
Global Positioning System Coordinates: 33° 39' 24.9" N, 113° 39' 20.3" W*
Geology: Jurassic Igneous Welded Tuff
U.S. Geological Survey 7.5 Minute Topographical Map: Hope

FROM INTERSTATE 10, BETWEEN TONOPAH AND QUARTZSITE, take exit 53 and proceed north on Hovatter Road. After going 2 miles, turn left (west) onto an unmarked gravel road. This is Hull Road. Follow Hull Rd. 3.1 miles where you will see a low ridge of light colored volcanic tuff on your right next to the north side of the road. If you are beyond Hope and coming from the north on U.S. 60, Hull Road leads south from U.S. 60 1 mile east of the intersection of U.S. 60 and State Route 72 in the little town of Hope. Follow Hull Road south 6.5 miles to the turnoff to collecting area. Turn north on the faint tracks that lead about 200 yards to the base of the ridge and park. From I-10 both Hovatter and Hull Roads are suitable for passenger cars. From U.S. 60, Hull road requires high clearance.

View of the collecting area from Hull Road.

Because chalcedony is a fairly common material found at several locations across Arizona, it holds little interest for most experienced collectors. Novice rockhounds, however, will find this location a friendly introduction to the hobby of mineral collecting. Unfortunately, the membership of most rock clubs in Arizona is elderly. Without an active recruiting program to attract younger people, especially children and young families, to the rockhound hobby, many clubs statewide are, like the old fossils we like to collect, in danger of becoming extinct. This site is a perfect club sponsored introductory field trip destination for children. The ground is flat and easy for little feet to walk on. There are no big rocks to get in the way, rough gullies and erosions to trip over, or mine shafts to fall into. And, there are no cactus or pricker bushes to puncture tender young skin. Near the base of the ridge, there are some small sloping tailings piles left by collectors who chiseled chalcedony specimens out of the tuff cliff. There is no need to climb on these cliffs, however, because the flat ground is littered with plenty of bright, shiny chalcedony for eager young ones to discover.

There is material of interest to experienced adult collectors as well. Although most chalcedony fluoresces, this chalcedony fluoresces as brilliant an emerald green as you will find anywhere. Along with the flat, curvy, chalcedony rose formations, there are also small triangular agate vugs, and little chalcedony and agate geodes about the size of a dime in the matrix and loose in the dirt. If you like to hammer and chisel, you can liberate large chalcedony plates and veins from the tuff.

*G.P.S. coordinates taken at the base of the ridge.

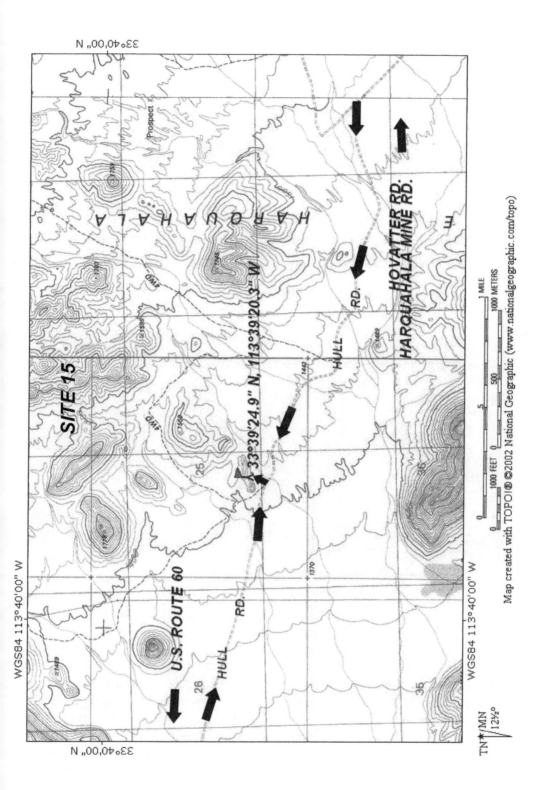

SITE 16

Malachite at the Mammoth Mine

Difficulty Scale: 7 – 5 – 3   Seasons: Fall, Winter, Spring
Global Positioning System Coordinates: 34° 02' 55.8" N, 113° 45' 53.3" W*
Geology: Middle-Early Proterozoic Various Granitic & Biotite Granite Rock
U.S. Geological Survey 7.5 Minute Topographical Map: Butler Pass

The road leading to the Mammoth Mine collecting area showing the abandoned trailer on the left.

FOLLOW ALAMO LAKE DAM ROAD 11.1 miles north from Wenden on U.S. 60. Turn left (west) on Transmission Line Road. Although this is a maintained gravel road all the way to Midway, passenger cars may experience some difficulty with loose sandy areas along the first half (south-east end) of the road. Drive 13.1 miles and turn right (north) on an un-maintained dirt road. Pay close attention as you navigate this road because there are severe washouts that are hidden from view until you are right on top of them. The detours around these

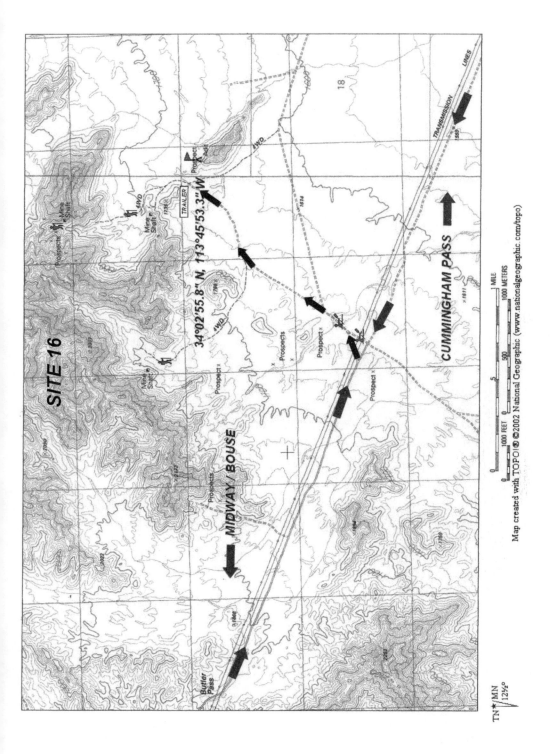

washouts require 4-wheel drive. As you approach this turnoff, you can see the remains of an abandoned house trailer in the hills at the collecting area on your right (north). If you are coming from Midway, the distance to the turnoff is 6.9 miles, 1 mile beyond Butler Pass. Follow the dirt road 1.5 miles. The collecting area is the excavation on the Hillside opposite the old house trailer. This mine has also gone by the names Chicago and Copper Glance Mines. There are additional mines within a mile of this location that may be worth exploring. But, since the roads leading to them have become obliterated, you will have to hike cross country to reach them.

Although small, the malachite specimens here are good. The material is hard, colorful, and gemmy. Look for rocks containing malachite veins and attempt to crack them open along the seams to retrieve malachite coated rock faces.

*G.P.S. coordinates taken at the collecting site.

SITE 17

Hematite and Copper Minerals at Cunningham Pass

Difficulty Scale: 5 – 5 – 3   Seasons: Fall, Winter, Spring
Global Positioning System Coordinates: 33° 57' 40.2" N, 133° 33' 40.8" W*
Geology: Early-Late Cretaceous Light Colored Granite and Pegmatites
U.S. Geological Survey 7.5 Minute Topographical Map: Cunningham Pass

The turnoff from Alamo Dam Road to the Cunningham Pass collecting areas.

FROM WENDEN ON U.S. ROUTE 60, drive north on Alamo Dam Road. Turn left (west) about 100 yards before reaching the 10 mile marker onto the unmaintained dirt road. As you make the turn, you will see three tailings piles. The one that is a few hundred yards directly ahead and to the left of you is the Little Giant Mine. You can park on the road directly beside it. The second location is on your left to the south, up the hill, directly beneath the power line. To reach it, follow the road beside the Little Giant tailings pile southward about .3

mile and park where you can turn around. From here, you will have to hike the last 200 – 300 feet to the collecting area. To reach the third collecting area, follow the road toward the tailings pile you see on the mountainside to the west. After driving about .8 mile, you will come to a road on your left leading south into a canyon. If you follow this road, you will find more excavations to explore. If you pass by it and continue westward, the Critic Mine will come into view on the mountainside on your left (south) within about .2 mile.

The mines in the Cunningham Pass are for collectors who are interested in acquiring a few small, but very attractive, specimens. No Large cabinet sized pieces here. Thumbnail size specimens are about the largest you can expect to find. The mines in this area were primarily copper prospects. You can find traces of malachite along with large amounts of hematite in the dumps. The best specimens are those displaying small, shiny malachite sprays, on a bright, black hematite background accompanied by delicate quartz crystal needles. These specimens are somewhere between micro-mounts and thumbnails in size. You don't need a microscope to see them, but they are best viewed through the magnification of a loop.

Turnoff to the Critic Mine

*G.P.S. coordinates taken at the intersection of the turnoff and Alamo Dam Road.

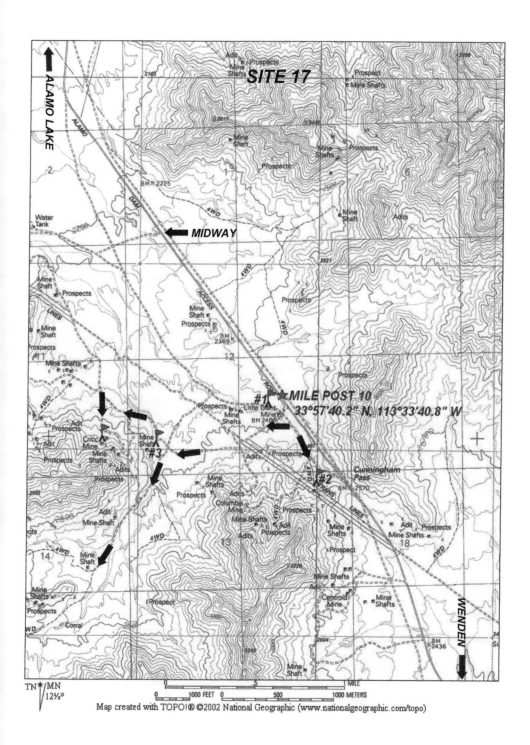

SITE 18

Copper Minerals at the Bullard Mine

Difficulty Scale: 3 – 5 – 3   Seasons: Fall, Winter, Spring

Global Positioning System Coordinates: 34° 03' 00.9" N, 113° 16' 43.8" W*

Geology: Early Proterozoic Metamorphic Schist and Gneiss

U.S. Geological Survey 7.5 Minute Topographical Map: Smith Peak

From the intersection of U.S. Route 60 and Eagle Eye Road in Aguila, drive 4.1 miles to mile post 31 and a sign that says "leaving Maricopa County." From here, follow the maintained road as it curves left around to the west. From the curve, drive 3.1 miles to an un-maintained road on your right leading northwest. It is marked with a sign saying "Smith Peak". Follow this road 2.8 miles to a fork at a large BLM signboard containing a map and information about the Smith Peak area. Take the left fork and drive 1.5 miles to the Bullard Mine access road on your right (north). Passenger cars can make it this far but, the roads leading to the collecting areas require high clearance. From this point, you can see the Bullard Mine excavations across the face of Bullard Mountain in front of you. You can see the tailings of the westernmost excavation at the end of the access road directly ahead of you. To the right, farther east on the mountainside, you will see the main part of the mine with its large tailings slide, wooden ore shoots, and road system. To reach this area, proceed up the access road and, within about 100 feet, turn right (east). After going .2 mile, you will see some shallow excavations on your right. There is nothing much to collect here. Immediately beyond these excavations, you will see a faint, narrow track leading off into the bushes on your left (north). Take this road .1 mile to a low mine dump containing chrysocolla beside a wash. From here, there is a road leading off to the east a few hundred yards that will return you to the road leading up to the main part of the mine.

The drive up to the westernmost area is .5 mile. You can park and turn around on top of the tailings plateau. Here, you can collect large chunks of dark colored calcite that fluoresces dull red. When struck, the large chunks will cleave into small, angular pieces revealing the calcite's trigonal crystal form. The most interesting specimens are those adorned with veins of chrysocolla. Along the short road leading eastward above the tailings plateau, you can collect

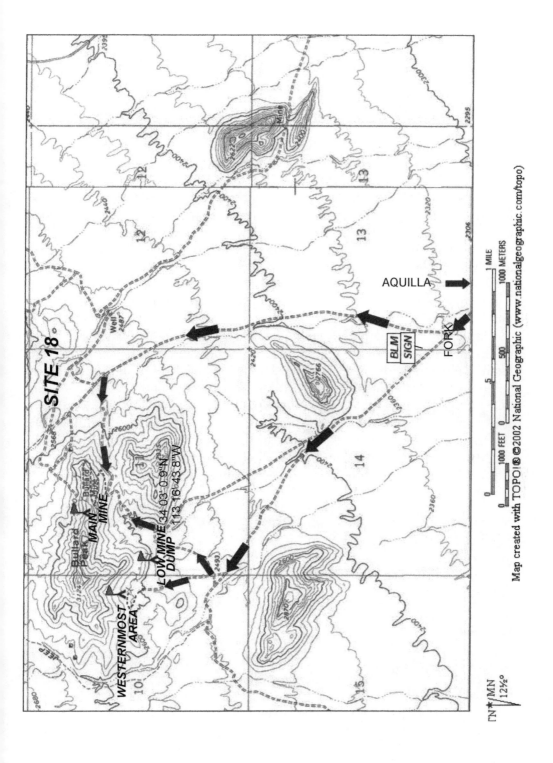

samples of light green epidote. If you walk to the end of this road, there will be a small adit below you beyond which you can see the low mine dump and the roads leading to it.

The best collecting location in the Bullard Mine area is the low mine dump. You will not be able to see it until you are upon it because it is so well hidden in the bushes. Lapidaries, especially those who fashion "Indian" jewelry, will find this location particularly interesting. The dump is a large pile of red dirt containing 2 – 6 inch rocks bearing gem quality chrysocolla and malachite. Although very colorful, the rock that is scattered around the dump has been pretty well picked over. Dig thought the soil to uncover the best material. Look for rocks containing wide blue and green veins that go through the rock from one side to the other. Often, these veins will be mixed with clear quartz and shiny black hematite matrix. In some cases, the dark brownish-red iron based host rock will be mottled with bright blobs of malachite and chrysocolla. Since the host rock is highly silicated, it will cut, grind, and tumble well and take a very bright polish. The patterns formed by the combination of all these minerals make very pretty gem stones.

At the main mine, there is an abundance of thin chrysocolla veneer in the dumps and on the walls of the cliffs surrounding the mine tunnel entrances. From the low mine dump, you may need 4-wheel drive for a short distance to get to the main mine area. Or, you can return to the fork at the BLM signboard, take the right fork (north) for 1.6 miles where the road on your left (west) will lead you .6 mile up to the main mine area. Search through the tailings for interesting, colorful, and unusual, chrysocolla specimens. If you follow the right fork around to the north side of Bullard Peak, you can explore more excavations if you wish to scramble up the mountainside to reach them.

View of the Bullard mine area from the turnoff.

G.P.S. coordinates taken at the low mine dump.

SITE 19

Amethyst at the Contact Mine

Difficulty Scale: 4 – 5 – 5   Seasons: Fall, Winter, Spring

Global Positioning System Coordinates: Given with each Location

Geology: Early Proterozoic Metamorphic Yavapai Supergroup & Pinal Schist

U. S. Geological Survey 7.5 Minute Topographical Maps:
Black Butte and Hummingbird Spring

FOLLOW THE DIRECTIONS ON PAGE 340 to the Vulture Mine–Wickenburg–Aguila Roads intersection. From the intersection, follow Aguila Road 11.2 miles and turn left (south) just after crossing Dead Horse Wash. This road is maintained in excellent condition to accommodate the large 18-wheel trucks that haul gravel from the huge Bighorn gravel pit 6 miles to the south. Follow this road south 3.7 miles and turn right (south) onto a faint, narrow road leading into a canyon. After about .2 mile, the road splits into a high road and a low road. Bear right onto the high road that curves up and around bypassing the washout on the low road. The high road then reconnects with the low road just before a second washout. You will have to park here and hike a short distance to two roads that lead up to two different collecting areas on the canyon's west slope. From the parking place, hike across the second washout and across a wash where the road begins to ascend the hillside. Follow the road upward a few hundred yards where you will come to the first road that leads upward through the ruins of an old mining camp, through a wash, and up a steep slope. After ascending about 200 feet, the road will make a 90° turn to the right (north), flatten out and run horizontally along the hillside. To the left of the turn (south) about 50 feet down the slope is a geodetic survey marker (G.P.S. coordinates 33° 44' 21" N, 113° 02' 16" W) This is the first collecting area. The second road leading up to the second collecting area is only a few hundred yards beyond (south of) the first road. Turn right (west) after passing the old rusty truck cab. The second road curves up the slope and then, like the first road, makes a right (north) turn and cuts horizontally across the hillside to two vertical mine shafts. The second collecting area is the road and the slope below the second mine shaft (G.P.S. coordinates 33° 44' 14" N, 113° 02' 11" W).

There is a considerable amount of float to collect on the roads and on the slopes below the roads at both locations. Amethyst chunks up to 6 inches across are available. Pieces that have been exposed to the sun look like ordinary quartz because they have faded. But, if you crack them open, you will find rich, gemmy, blue amethyst color. At the first location, you can, with a heavy hammer and pry bar, extract material directly from the uphill ledge created by the road cut. Look for crystally vugs and veins up to 3 inches thick. At the second location, look for pieces containing malachite and chrysocolla in conjunction with the amethyst. These were copper mines that penetrated an amethyst bearing stratum as the mine shafts were sunk.

The first road leading up through the ruins of the old mining camp.

View of the second road curving up the slope to the second collecting location.

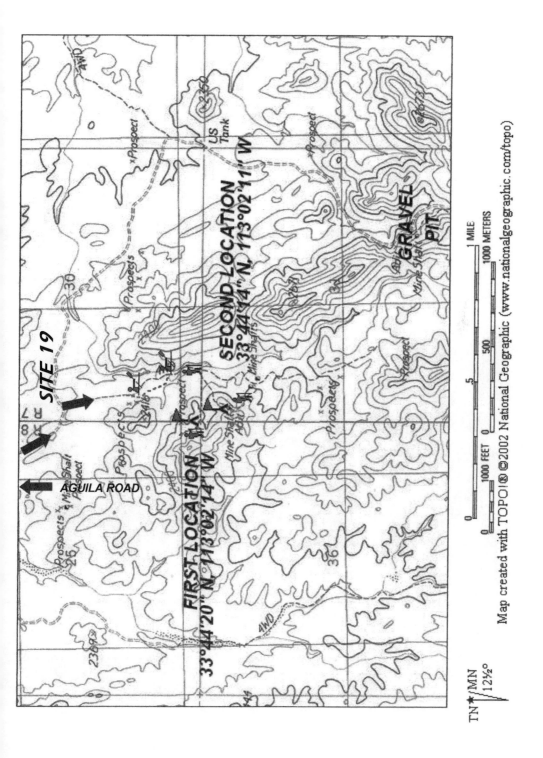

SITE 20

Fluorite and Selenite off Eagle Eye Road

Difficulty Scale: 4 – 6 – 6   Seasons: Fall, Winter, Spring
Global Positioning System Coordinates: 33° 42' 46" N, 110° 18' 07.4" W*
Geology: Middle Miocene-Oligocene Sedimentary & Volcanic Tuff & Lava
U.S. Geological Survey 7.5 Minute Topographical Map: Weldon Hill

The faint track leading off into the bushes to the collecting area.

FROM INTERSTATE 10, TAKE EXIT 81, Buckeye-Salome Road. Drive north-west 9.7 miles toward Salome and turn right (north-east) on Eagle Eye Road. Drive 7.6 miles where you will see a small mountain on your right. From Aguila, take Eagle Eye Road south 20 miles. The collecting area is a short road-cut about one-quarter the way up the mountain side. The narrow road up to the collecting area is on your right (south-west). It is difficult to spot because it is only a faint track leading off into the bushes. Directly across Eagle Eye Road on your left is a more visible dirt road leading off to the north-west. You may have to park and search for the road up to the collecting site. Once you have found it, drive in about 100 yards to two wooden stakes marking a geodetic survey marker. Follow the tracks left at the survey marker about .1 mile and park. It is a short but rather steep hike up to the collecting area. Do not attempt to drive up to the collecting area. There is not enough room for even an ATV to turn around up there. The collecting area is the uphill banking of the cut where the road levels out and goes horizontally about 100 yards across the face of the hill.

This site, known as the Lemon Fluorspar Prospect, is an excellent destination for kids between the ages of about 10 and whenever they decide that it is no longer cool to be seen with their parents. Small, narrow bands of fluorite run horizontally in the banking beside the road ready to be discovered and harvested by eager young miners. The banking is composed of rather colorful limestone, dirt, and fibrous snow-white selenite that is so delicate that it turns to powder if not handled extra carefully. Digging into this bank to retrieve the fluorite and selenite is moderately easy. In addition to the standard rock pick, you will need a 3 pound hammer, a chisel or two, and a small pick. There are no museum quality specimens here. But, there are shiny, emerald green fluorites lurking in the dirt ready to be unearthed. It is a great opportunity for kids to learn that rewards can be had with a little hard work. On the ground, you will see small pieces of clear fluorite that have faded in the sun. Dig where others have dug before to find fresh, colorful material.

The spooky Black Rock Mine quarry.

Up the road from the fluorite site, in the Big Horn Mountains, is another interesting area for both kids and adults to visit. Continue up Eagle Eye Road another 5.9 miles and turn right (south-west) on Little Horn Peak Road. The area around Little Horn Peak is a former manganese mining district. Although there is nothing worth collecting here, you can gain an understanding of what large scale mining is like. Follow Little Horn Peak Road 3.8 miles to Little Horn Peak. Here, the road curves left, to the west, past a large tailings dump and brings you to the edge of the huge open pit known as the Black Rock or Black Warrior Mine (33° 44' 10" N, 113° 11' 03" W). You can follow the road down into the pit where you can park and turn around.

This quarry is a very spooky place. As you descend into its depths past the rusty old machinery, the rock walls get black as coal and equally as dirty. Manganese ore is soft, grainy, ugly, dusty, and soils everything that it touches. In a 180° arc from east to west, where the 200 foot vertical sides met the quarry floor, a dozen or more large gaping tunnels penetrate the quarry walls. These excavations go off in several different directions forming a labyrinth of mysterious chambers and interconnecting passageways like the dungeons in a gothic, medieval castle. Abandon hope all ye who enter here!

*G.P.S. coordinates taken at the intersection of Eagle Eye Road and the turnoff to the site.

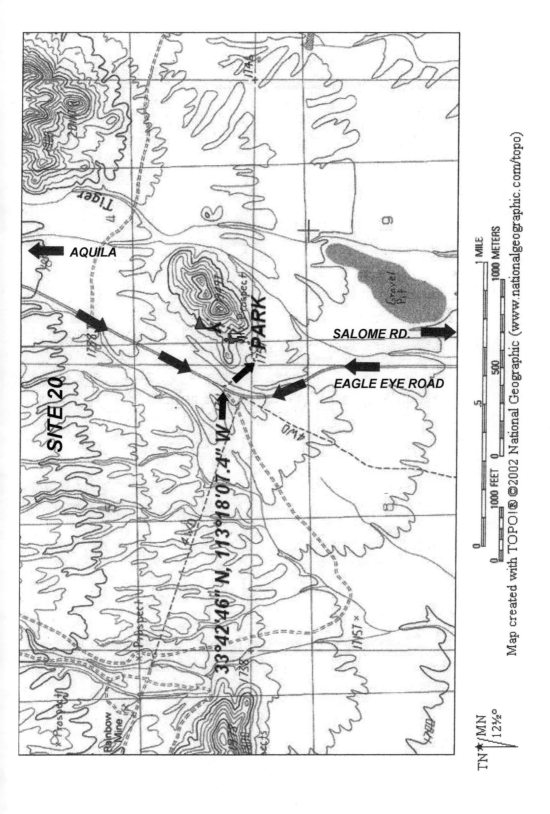

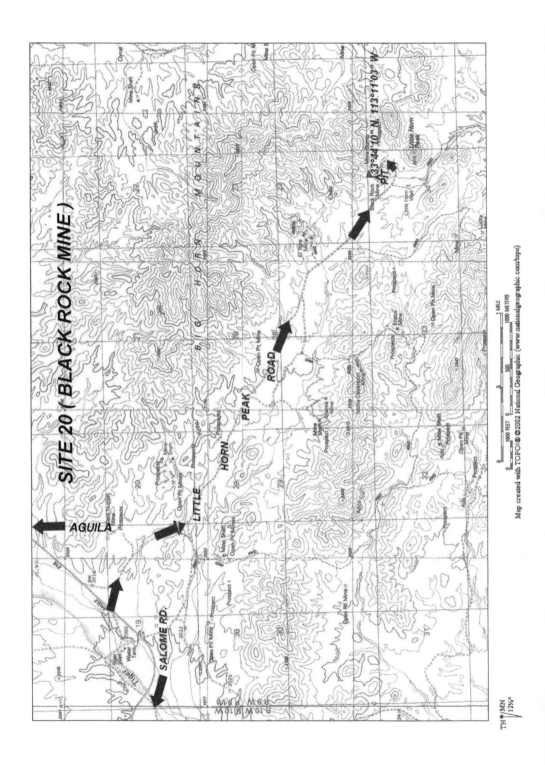

SITE 24

Jasper and Hematite at Mingus Mountain

Difficulty Scale: 3 – 3 – 5   Seasons: Spring, Summer, Fall

Global Positioning System Coordinates: 34° 40' 08.1" N, 112° 09' 08.5" W*

Geology: Early Proterozoic Metamorphic Schist and Gneiss

U.S. Geological Survey 7.5 Minute Topographical Map: Hickey Mountain

TO REACH THIS SITE, turn off State Route 89A between mile markers 336 and 337 into the Mingus Mountain Recreation Area. The distance from the Jerome Fire Station is 7.7 miles and about 23 miles from Prescott. From the entrance to the Mingus Mountain Recreation Area, follow the road eastward 1.5 miles and turn right (south). Go 2.2 miles and turn right (south) on Forest Road 132. Drive .4 mile to the intersection of F.R. 132 and F.R.105. Park in the open area by the water hole. The collecting area is the forested hillside on your left as you face the water hole.

Walk up the hillside through the trees a few hundred yards. The G.P.S. coordinates above are at the point where you will begin to encounter float material. As you continue up the slope, you will find outcrops of material running up and down the hillside. There are some pieces left on the surface from previous digs that you can collect. If you are more energetic, you can dig and pry out larger chunks from the outcrops. You will have to work carefully, because the rock tends to separate in sheets between the layers. If you do any digging, remember to restore the site to its original condition. Do not disturb any trees or bushes. Be sure to fill in any holes you made to prevent erosion. Failure to properly maintain the site could possibly result in the Forest Service closing it to future collecting.

Jasper, quartz, and hematite are fairly common and generally unremarkable substances. Together at this location, however, they form an interesting lapidary rock. The jasper is a medium to dark red color. The hematite is dense, solid, and hard. It ranges in color from a bright metallic jet black to a light rusty red. These substances are arranged in alternating layers several inches thick.

Intermittently, a layer of clear quartz interrupts the layering sequence. To create striped slabs, cut the material vertically 90° to the horizontal. These slabs can then be fashioned into cabochons and other lapidary forms.

The collecting area on the hillside beside the water hole.

*G.P.S. coordinates taken at beginning of collecting area.

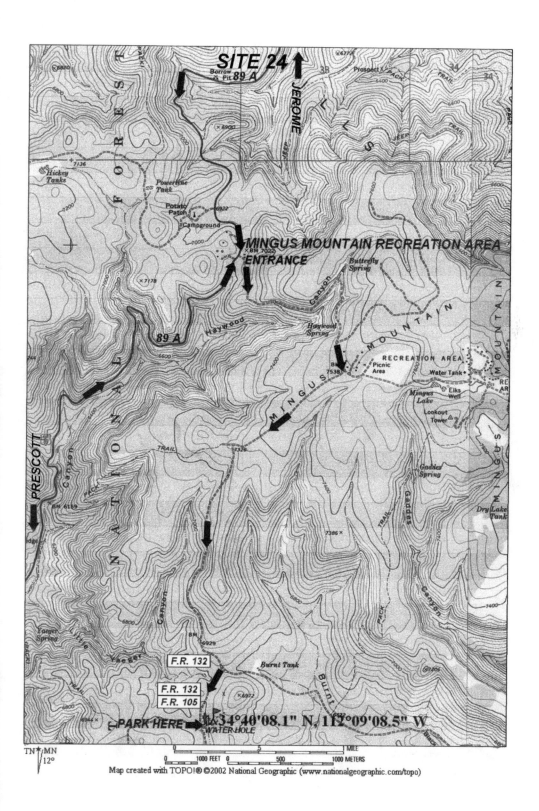

SITE 25

Siderite at the Copper Chief Mine

Difficulty Scale: 4 – 4 – 3   Seasons: All

Global Positioning System Coordinates: 34° 42' 11" N, 112° 05' 38" W

Geology: Pliocene-Middle Miocene Rhyolitic and Andesitic Lava

U.S. Geological Survey 7.5 Minute Topographical Map: Cottonwood

FROM THE INTERSECTION of Main Street and U.S. Route 89A in Cottonwood, go south on 89A, toward Prescott, 1.7 miles to Mingus Avenue. Turn left (south) on Mingus Ave. and drive 3 miles to a fork. After driving .8 mile the pavement ends at a cattle guard. Bear right at the fork on Forest Road 493. Follow F.R. 493 for 2.5 miles to the turnoff to the Copper Chief Mine on your left. As you round a curve a few tenths of a mile before reaching the turnoff to the collecting area,

View of the Copper Chief Mine as you round the curve on Forest Road 493.

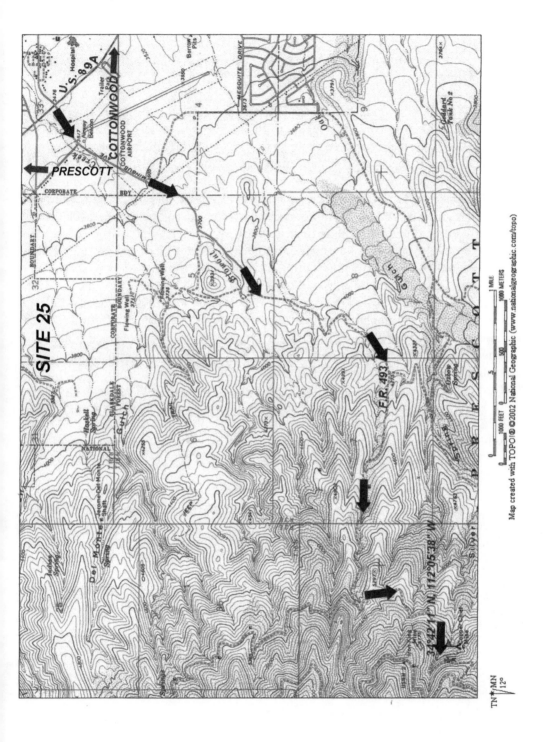

you will see the tailings slide and concrete foundations of the Copper Chief Mine on the mountainside ahead of you. The road into the mine is blocked so you will have to park and walk approximately .2 mile to the top of the mine dump.

The collecting area consists of the tailings slide, the tailings plateau, and the graded hillside above the plateau. Roam the area looking for bright milky white bull quartz studded with massive dark metallic and mahogany colored siderite. Siderite is an iron carbonate and a member of the calcite family. Consequently, it looks rather like massive calcite chunks with an iridescent metallic luster. Exposed surfaces are dark and have a metallic luster whereas, internal surfaces display lighter color and a more pearly luster when fractured. Specimens are easy to find as they shine in the sunlight like pieces of highly polished metal scattered among the rubble and vegetation on the ground. If you are not satisfied with the abundant float that litters the surface, you can easily dig into the soft, loose tailings and bankings in search of larger specimens. Most of the float pieces are about baseball size or smaller and display the crystalline architecture of similar sized chunks of common calcite. Closer inspection reveals trigonal crystal formations and, where fractured, perfect cleavage plains. A chevron crystal habit is evident on some specimens. Color varies from a light milk chocolate brown to a deep gun metal blue. When light reflects from it at the proper angle, the mineral displays a fiery iridescence much like specularite. Most of the siderite here is attached to or forms veins coursing through pure white bull quartz. Together, the dark, shiny, metallic specularite and the bright white quartz make a striking combination.

SITE 26

Travertine at the Empire Onyx Quarries

Difficulty Scale: 6 – 4 – 3   Seasons: Fall, Winter, Spring
Global Positioning System Coordinates: 34° 39' 31" N, 111° 45' 12" W*
Geology: Pliocene-Middle Miocene Sedimentary Limestone
U.S. Geological Survey 7.5 Minute Topographical Map: Cosner Butte

Quarry #1, the north quarry.

FROM INTERSTATE 17, TAKE EXIT 298 and turn east onto Forest Road 618. From this same exit, State Route 179 leads westward to Sedona. Travel east .5 mile on F.R. 618 and turn right (south) on F.R. 119. Drive 2 miles on F.R. 119 and turn right (west) onto the narrow unmarked road that leads around and up the low bluff where the quarries are located. Follow this road .5 mile to the rough little road that leads uphill to quarry number 1, the north quarry. You may need four

Quarry#2, the south quarry.

wheel drive to get beyond a steep, eroded, ten yard stretch of road. If you have to park and walk, the distance to the quarry is only .25 mile.

This site is composed of two quarries yielding massive, clear, white, and light brown, banded travertine. Travertine is one of several forms of calcium carbonate. Varieties include sedimentary limestone and chalk, metamorphic marble, and hydrothermal travertine. As water containing dissolved calcium and carbon ions percolates through the soil or evaporates on the surface, the calcium carbonate comes out of solution and crystallizes forming layers of vitreous, gemmy travertine. Over time, these layers can become several feet thick. Depending upon what other minerals may be present at different times, these layers can develop very attractive multi-colored bands. Because it is lustrous, colorful, abundant, and easy to work, it has long been a popular medium for sculptors and lapidaries.

Of the two quarries, number 1, the north quarry, is the most suitable for the average collector. The area has been graded leaving flat zones as well as piles of soil and rock forming a swath about 100 yards wide by about 300 yards long leading up a gentle slope. Small chunks of travertine measuring only a few inches across are visible and easily collected in the graded soil. If you are a little more energetic, you can dig into the dirt and rock piles in search of larger pieces. If you enjoy heavy manual labor, this location offers two opportunities. First, there are several piles of thick, large, travertine bearing slabs available. Some of these display wide subtitely banded layers of white and light-brown travertine that can only be harvested with a sledge hammer. Or, if you are

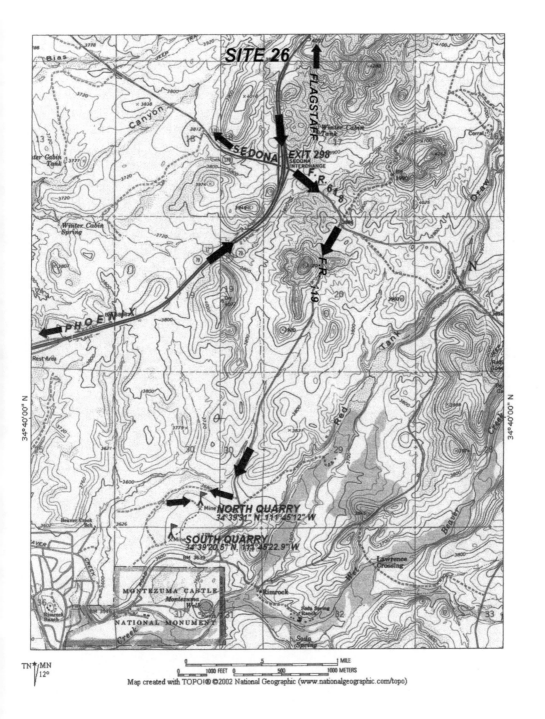

determined and patient, you can attempt to work the low ledge that is exposed up the slope at the south end of the quarry. In this ledge, a gemmy substratum of clear and whitish translucent travertine about six inches thick is visible beneath a 2 – 3 foot common limestone overburden. You will need a heavy sledge and large wide blade chisels to remove useful size chunks intact.

Quarry number 2, the south quarry, is .3 miles farther down the road from the north quarry. This excavation is more like you expect a stone quarry to look like. The rock ledge is geometrically cut into squares, rectangles and flat planes forming stair step terraces from top to bottom. Very large limestone blocks, some containing zones of travertine rest around the rim of the quarry. These are interesting to see, but obviously much too big to collect. There is, however, a lot of smaller sized quarry rubble on the slope overlooking F.R. 119 at the southeast end of the quarry. Search these tailings for crystalline, drusey, vuggy specimens. Search carefully, because the tailings slope is steep and loose.

G.P.S. coordinates taken at quarry #1.

SITE 27

Glauberite Pseudomorphs along Salt Mine Road

Difficulty Scale: 3 – 4 – 4   Seasons: Fall, Winter, Spring

Global Positioning System Coordinates: Given with each location

Geology: Pliocene-Middle Miocene Sedimentary Limestone, Mudstone, Gypsum

U.S. Geological Survey 7.5 Topographical Maps: Camp Verde and Middle Verde

Location #1 G.P.S. Coordinates: 34° 32' 19.9" N, 111° 53' 32.7" W*

Location # 1, overlooking the canyon from the parking area.

FROM INTERSTATE 17, TAKE EXIT 285, General Crook Trail on the south side of the Verde Valley. After going about 1.8 miles, you will come to a stop sign. Turn right (east), go .2 mile, and turn right (south) on Oasis Road. Drive .45 mile down the hill to a dirt road on your right just before the intersection of Oasis Road and Salt Mine Road. Turn right (west) onto the dirt road. This road is rutty and un-maintained requiring high clearance. Drive .9 mile to a cattle guard, bear left, and drive .65 mile across a wash and up a hill to a turnout on the left hand side of the road overlooking a canyon and wash below. Park here and scramble down the faint trail to the streambed. This canyon is much like the one at location #2 only deeper. When you reach the bottom, walk downstream (north) a few hundred feet where you will discover calcite and aragonite after glauberite crystal clusters. There are many more calcites than aragonites. However, it is worth the effort to search out the aragonite clusters because their bright golden color gives a gemmy appearance unlike the dull earthy looking calcite crystals. Both varieties are found in the stream bed and the bankings.

Location #2, Ryal Canyon, G.P.S. Coordinates:
34° 31' 43.4" N, 11° 52' 05.4" W*

Location #2, Ryal Canyon, from the turnout on Salt Mine Road.

Return to the intersection of Oasis Road and Salt Mine Road. Bear right (south) on Salt Mine Road and follow it for 3 miles to a turnout on the right. The road is paved all the way to the turnout. This is location #2 at Ryal Canyon. The collecting area is about 500 yards to the south in the wash between the telephone poles. Walk down the short banking from the parking area and follow the dry streambed southward for about a quarter of a mile. The streambed will enter a narrow little canyon where the bankings become steeper and higher. From the beginning, you will encounter white selenite after glauberite crystal clusters. They are even in the gutters and road cut bankings of Salt Mine Road. Keep your eye out for clear selenite clusters in the streambed and in the muddy canyon walls just before you get to the waterfall. Above the waterfall, which is a three-foot step-up in the streambed, you will find larger gypsum-white crystal clusters. These are more square and blocky than the clear ones that are thinner, tapered, and sharp edged. Occasionally, both types will occur together in the same cluster. Search for clusters in the streambed and in the canyon walls. You can simply pick them up or tease them out of the soil with your rock pick. Be

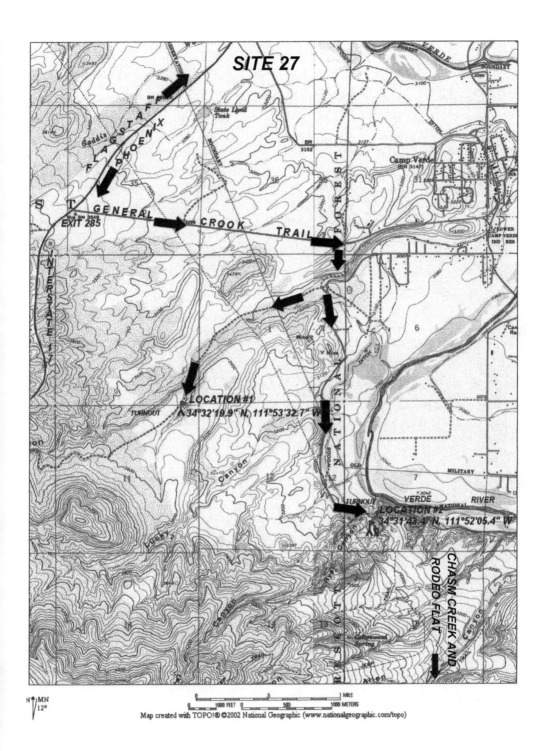

selective. Look for those that have several blades fanning out from the center in several directions like the Sydney Opera House. The clusters are usually very dirty with dried mud packed between the blades. The dirt and mud flake off rather easily. Then, you can wash off the rest with a toothbrush.

The term pseudomorph is Greek *pseudomorphos* meaning false form. Mineralogically speaking, a pseudomorph is a mineral that displays the outward shape or form of another mineral. The calcite, aragonite, and selenite after glauberite pseudomorphs described in this chapter are formed by a three step replacement process. First, glauberite crystals are formed in the soft Verde Limestone Formation sedimentary silts, clays, and salt deposits common to this area. Second, the glauberite is dissolved leaving an empty mold in the matrix in the shape of a glauberite crystal. Finally, the empty molds fill up with calcite, aragonite, or selenite. When the matrix wears away the replacement minerals in the shape of glauberite are revealed.

*G.P.S. coordinates are taken at the parking turnouts.

SITE 29

Calcite near Forest Road 41

Difficulty Scale: 5 – 6 – 5   Seasons: Fall, Winter, Spring

Global Positioning System Coordinates: 33° 58' 17.9" N, 112° 03' 50.1" W*

Geology: Middle Miocene-Oligocene Volcanic and Sedimentary Rock

U.S. Geological Survey 7.5 Minute Topographical Map: Daisy Mountain

FROM INTERSTATE 17 between New River and Black Canyon City, take exit 236 and turn east on Table Mesa Road. Go 1 mile to a gate marking the beginning of Forest Road 41. Go 3.3 miles to another gate and cattle guard. You will drive through a working stone quarry on the way. Proceed .5 mile beyond the gate and turn right (south) on the little, narrow, dirt road. Follow it .1 mile to where the road ends in front of a 20 foot cliff which is a waterfall during the rainy season. Park and scramble up the slope on your left (east) to the top of the cliff. From here, follow the wash up into the canyon. Within the first 100 yards, you will begin to see material eroding out of the bankings and side gullies.

The second gate on Forest Road 41 .5 mile before the turnoff to the collecting area.

Approximately 200 – 300 yards into the canyon, you will come to a wide amphitheatre-like opening with steep slopes and gullies on all sides. The slope on the right (west) as you enter the amphitheatre contains small, 1 to 3 inch across, rocks that are vugs that have weathered out of the soft, crumbly igneous flows. The rocks are composed of chalcedony, agate, and calcite in attractive patterns and color combinations. These vugs are generally triangular or rectangular in shape. The Chalcedony fluoresces a pretty bright green and some of the calcite fluoresces a mellow pinkish-red. Together, the minerals in these vugs make a pleasing fluorescent display.

Farther into the Amphitheatre, you will find plates of angel-wing calcite weathering out of the dusty, grainy, limestone slopes on your left (east). Look in the gullies, the wash, and on the ground below the slopes for specimens. Although these plates may be as much as 2 inches thick, they require careful handling because they tend to be fragile and crumbly. Bring adequate wrapping material in which to pack them out. In the slopes above, you can see ribs of calcite protruding from the soil. If you choose to climb these slopes to harvest this material, be careful. The ground is very slippery. Although it looks like you can sink your feet into the granular soil, only the first inch or two is soft. It is too hard underneath to gain a foothold. You can find more material if you follow the wash farther up the canyon.

G.P.S. coordinates taken at parking place.

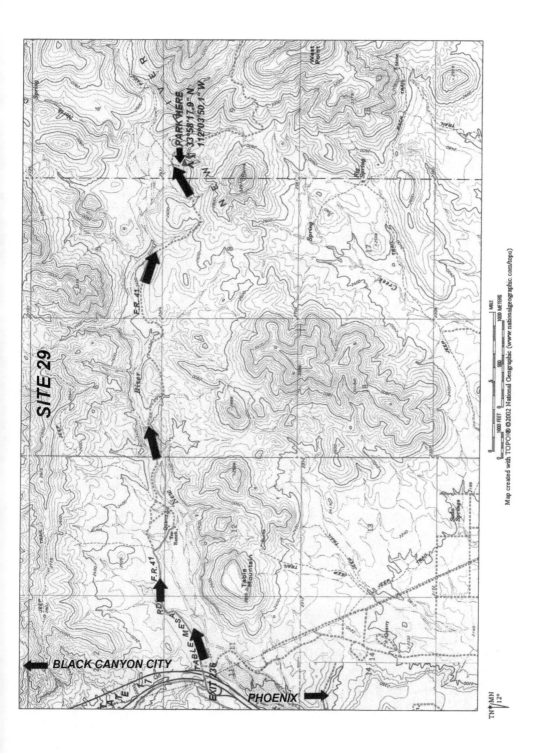

SITE 30

Banded Jasper off Table Mesa Road

Difficulty Scale: 2 – 4 – 4   Seasons: Fall, Winter, Spring

Global Positioning System Coordinates: 33° 59' 10.5" N, 112° 10' 34.6" W*

Geology: Pliocene-Middle Miocene Igneous Rhyolite and Andesite Flows

U.S. Geological Survey 7.5 Topographical Map: New River

FROM INTERSTATE 17 BETWEEN New River and Black Canyon City, take exit 236, Table Mesa Road. At the west end of the overpass above Interstate 17, turn north at the Frontage Road sign. The "frontage road" will parallel the freeway for only a short distance before turning north-west toward the Bradshaw Mountains. From the I-17 exit, go 1.7 miles to a fork in the road. Bear left and go 2 miles. Turn right on the road under the power poles. Drive down this road .15 mile and park at the base of the hills. The collecting area is the cluster of 4 or 5 hilltops with the rocky outcrops on your left (west). Be careful here. This area is obviously a popular shooting range. The parking area is littered with shell casings of every caliber. It might be advisable to park in front of the banking where targets are placed so that your presence in the area is known to shooters who

The road under the power poles leading to the site.

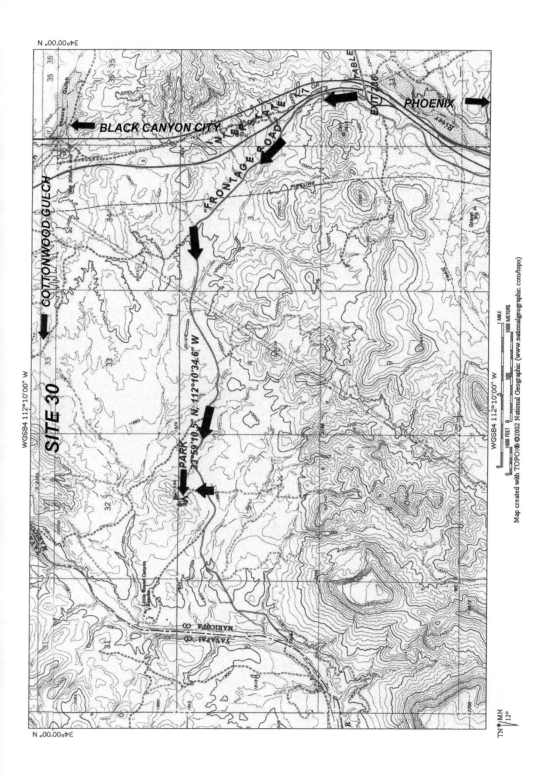

may come along. The collecting hills are directly in the line of fire. It is possible to drive up onto the hilltops, but there is really no need to do so.

This site offers some excellent lapidary material in the form of hard, dense, large, banded jasper rocks. The rock is beautifully striped and colored in the kind of rich pigments you would expect to find on a fine artist's palate. Blues and grays, reds and browns, tans and creams, oranges and yellows, and other subtle tints and shades alternate randomly in strips as wide as one quarter of an inch and as narrow as a pencil line. Some pieces display more or less parallel stripes. Others resemble the bent and twisted contours of petrified wood roots and burls. Although most of the material is dense and thick enough for making large, wide slabs, some pieces contain chalcedony and calcite filled vugs that fluoresce nicely.

It looks as if this area was prospected with a bulldozer. There are large boulders as well as piles of smaller broken up rocks available. You can easily pick up workable pieces for cutting and polishing. Or, if you really want to work, you can try breaking the bigger boulders up with a sledge. These rocks will fracture randomly without the banded layers exfoliating.

G.P.S. coordinates taken at parking area.

SITE 32

Mica at Cottonwood Gulch

Difficulty Scale: 5 – 4 – 6   Seasons: Fall, Winter, Spring
Global Positioning System Coordinates: 34° 02' 03.2" N, 112° 12' 25.6" W*
Geology: Pliocene-Middle Miocene Volcanic Basalt, Tuff, Agglomerate
U.S. Geological Survey 7.5 Minute Topographical Map: Black Canyon City

View of the quartz reef valley from the road.

FOLLOW THE DIRECTIONS TO SITE 31 on page 333. Continue driving west past Site 31 .7 mile where you will come to a fork in the road at a corral. Bear left on the rougher road that goes downhill into a little wash. Follow this road for another mile to a small turnout on a hillside beside a quartz outcrop on the right (west) side of the road. The total distance is 1.7 miles from Site 31. Down in the valley on your right, you will see a quartz reef paralleling the road. The entire length of the reef is not visible. Only outcrops appear intermittently in a line for several hundred yards like the vertebra of some giant prehistoric serpent. You can collect at the outcrop beside the road or hike down the slope to the ones below. The walk is fairly easy and the distance is not too far.

The geology between Interstate 17 and Site 31 is primarily igneous with some limestone sediments here and there. By the time you reach this site, you have transitioned into a zone containing large quartz/mica pegmatites. The exterior surfaces of these quartz outcrops are weathered and discolored by dirt and organic growths. When broken, however, the interior surfaces are bright and shiny. You will need a heavy hammer to split the resilient quartz matrix apart. The quartz is snow white and lustrous. Most of the mica is silver colored but some has a greenish hue. The mica here is generally opaque and metallic looking. When you break up the quartz you may expose some good examples of mica books as well as chunky looking and drusey occurrences. Occasionally, some black schorl appears along with the mica adding interest and color to your specimen. Unfortunately, there is no tourmaline or other gemmy pegmatite mineral crystals here.

*G.P.S. coordinates taken at the turnout.

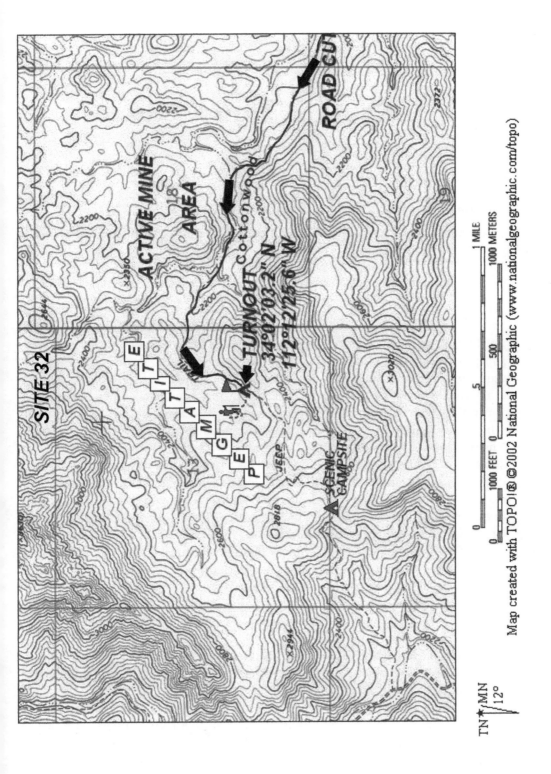

SITE 33

Agate at Signal City

Difficulty Scale: 4 – 5 – 5   Seasons: Fall, Winter, Spring
Global Positioning System Coordinates: 34° 28' 21" N, 113° 37' 36" W*
Geology: Early Proterozoic Granitic Varieties, Gneiss, and Biotite Granite
U.S. Geological Survey 7.5 Minute Topographical Map: Signal

THIS SITE AS WELL AS SITES 34 through 37 is located in a particularly remote and isolated area. The nearest civilization is a place called, appropriately, "Nothing". Entrance to and exit from the area is limited. Access from the north, east, and south is from U.S. Route 93 via Signal or 17 Mile Roads. Both of these routes require fording the Big Sandy River, once from 17 Mile Road and three times along Signal Road. Crossing is impossible during rainy periods when the river is running. From the north-west, access is from Yucca on Alamo Crossing Road which goes south to Alamo Lake. Despite its name, however, there is no crossing. The road dead ends at Alamo lake making access to and from the area from the south impossible. Therefore, plan your trips to these sites carefully. Bring a few days' extra food and water with you in case of emergency. Be sure you have enough gasoline to make it back to civilization even if you unexpectedly have to exit by the longest route available. From the intersection of Signal and Alamo Crossing Roads, the distance to Nothing on U.S. Route 93 via 17 Mile Road is 28 miles. The drive to Wikieup via Signal Road is 26 miles. If the Big Sandy fords are flooded, the alternate route back to Wikieup is 18 miles long via Alamo Crossing and Chicken Springs Roads. And, Yucca via Alamo Crossing Road is 40 miles away.

From U.S. Route 93, 66 miles north of Wickenburg, turn left (south) on 17 Mile Road. 17 Mile Road only goes 14 miles—no one seems to know why—where it intersects with Signal Road. Turn left onto Signal Road. Within .4 mile, you will arrive at the Big Sandy River Ford Number 3. If you are traveling from Wikieup, drive south on U.S. 93 about 8.5 miles and turn right (south-west) on Signal Road. Follow Signal Road 8.8 miles to Ford Number 3. Coming from this direction, you will ford the Big Sandy twice before arriving at Ford Number 3. From the point where you enter Ford Number 3, drive 3.7 miles to the entrance to Signal on your left (east). As you turn in, you will be able to see a mine on the

View of the ridge and mine workings at Signal.

east end of the little ridge on your left. Follow the roads that lead to the bottom of the ridge below the mine about .2 mile, park, and walk up to the collecting area.

This site features two collectables. First, there is silicated rock containing bands of agate, clear quartz crystal, and amethyst. Second, there is multi-colored fluorescent material. The best collecting is at the upper levels of the mine near the top of the ridge. Look for the dark blackish-brown colored rocks. There are a lot of small pieces, 2 – 5 inches across to select from if you do not want to work very hard. Unfortunately, exposure to the sun has caused any amethyst in these small pieces to fade. If you do not mind swinging a sledge hammer, you can liberate the freshest and most colorful material from the larger boulders that lay scattered around the mine site. Look among the dark colored rocks for those bearing calcite. These pieces fluoresce bright red (calcite) as well as green (chalcedony?) and blue-white (scheelite?).

Signal City, know only as Signal today, was the mill site for processing the silver ore from the McCraken Mine 9 miles to the west (see Site 34, page 128). When the McCraken Mine opened in 1874, the nearest settlement was Greenwood City upstream from what became Signal City on the eastern shore of the Big Sandy River. As the mines upstream played out by 1876, whatever supplies and equipment that could be moved was transported south to a new location on the western shore of the river. Hence, Greenwood City died and Signal City was born. By 1878, 2 mills, the Signal Smelter Mill and the McCraken Stamp Mill, employed several hundred men and were working around the clock.

The population of the town was about 1,000 people. There were 5 general merchandize stores, 3 restaurants, 13 saloons, and several houses, barns, and warehouses. Buy 1890 the McCraken mine had shut down and the census that year reported that the population had shrunk to only 80 people. In the years that followed, attempts to reopen the mine were unprofitable and both the mine and the town were abandoned.

G.P.S. Coordinates taken at the top of the ridge.

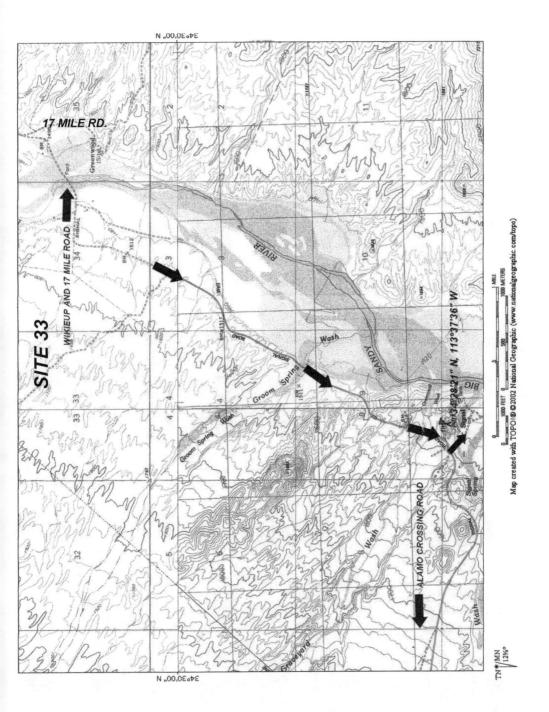

SITE 34

Banded Agate at the McCracken Mine

Difficulty Scale: 5 – 5 – 6   Seasons: Fall, Winter, Spring

Global Positioning System Coordinates: 34° 27' 04.8" N, 113° 46' 21.2" W*

Geology: Middle-Early Proterozoic Igneous Granitic Rocks

U.S. Geological Survey 7.5 Minute Topographical Map: McCraken Peak

FROM U.S. ROUTE 93, turn west on Signal Road, which intersects U.S. Route 93 8.55 miles south of Chicken Springs Road in Wikieup. Travel west 18 miles crossing the Big Sandy River three times until you reach the four-cornered intersection of Signal Road and Alamo Crossing Road. If you are coming from Wickenburg, follow the directions in Site 33, page 124. At this point, Signal Road becomes McCracken Mine Road (unmarked) as it crosses Alamo Crossing Road and continues west 3.6 miles to the McCraken Mine. You can see McCraken Peak to the west from the intersection.

The intersection of Signal and Alamo Crossing Roads showing McCraken Peak in the background.

Jackson McCraken was born in South Carolina in 1821 and migrated to Arizona in 1861. On August 17, 1874, he discovered silver on the mountain that now bears his name. He was one of 12 miners elected to the 27 member first Arizona Territorial Legislature. However, fellow members voted to unseat him because he refused to bathe. So, they dragged him out of town and submerged him in Groom Creek. It could have been worse, usually in the old West when a mob dragged you out of town, it was not for a bath. The McCraken was one of the most well-known and productive silver mines in Arizona. Its ore assayed at between $60.00 and $600.00 a ton. By 1880, the mine had produced over 6 million dollars.

McCracken road takes you directly to the foot of McCracken Mountain where it ends on a large, flat, graded pad with a concrete loading platform at the east end and two roads leading up into the mountain at the west end. This pad makes a good camping site and staging area for exploring the extensive excavations that cover the entire mountain from top to bottom. At the west end of the pad, the road to the right leads up to a ridge line, goes over a little pass, and down the west side of McCraken Mountain where you will find tailings, ruins, and the concrete foundations of the mine's ore processing plant. The dumps here contain interesting calcite, quartz, and agate formations. The terrain is steep and the ground is littered with sheet metal, sharp objects, and boards with protruding rusty nails. The road to the left winds its way up to several excavation levels on the northeast side of the mountain. This road is badly eroded at the beginning requiring 4-wheel drive. At the first level, it intersects the road coming up from the ore processing plant area. It then bears left and upward in an easterly direction winding its way up to the quarries above. All of the roads in the area require high clearance and four wheel drive in a few places.

There is collectable material everywhere; in the quarries and benches, in the erosions and bankings along the roadsides, and in the trench walls and

Large boulders containing wide veins of wavy banded agate.

cliffs. Therefore, the best approach is to survey the entire collecting area first avoiding the temptation to keep anything until you have seen everything. Then, you can concentrate on highgrading the best spots. There is literally a whole mountain of material here. Collecting is as easy or as hard as you care to make it. Collectable rocks range in size from boulders to pebbles. You can search the ground for cabinet sized specimens or, if you enjoy swinging a sledgehammer, you can attempt to bust up the larger boulders. Digging and prying into the cliff faces and walls will also yield pretty crystal vugs and seams.

McCraken Mountain is laced with huge wide veins of wavy banded quartz, some as much as ten feet wide. These bands vary in color and type. There are bands of solid white quartz, clear quartz crystal, blue amethyst, reddish and rose colored agate, zones of black and reddish-brown hematite, orange-yellow carnelian and citrine, and layers of course silica. The best veins contain all of the above. Patterns vary widely. Some veins have strait horizontal banding and others present a wavy sine-wave appearance. Still others appear to be completely chaotic with no predictable or repetitive pattern at all. Color sequencing varies as much as patterning. All this variety presents a wealth of opportunity for lapidary artists. The material is hard and dense, cuts well and takes an excellent polish, and even comes in pieces massive enough to make into tabletops.

For those who prefer natural unmachined specimens, the quarry walls contain vugs of quartz, calcite, and occasionally selenite. The matrix containing these vugs is hard and will break cleanly when hammered or chiseled. Vugs are clean, wavy, and meander throughout the matrix following no particular pattern. Some are round and cavernous and others are long and narrow. They are lined with short, squatty, clear, brilliant quartz crystals. Perched on top of the quartz crystals are clear bright angel-wing calcite crystals. Search among the rubble piles at the base of the quarry walls and cliffs for rocks containing these bright gemmy vugs. Quartz crystal vugs are not rare and are not always particularly attractive. But, these are exceptionally pretty and worth collecting.

*G.P.S. coordinates taken at the west end of the flat graded pad.

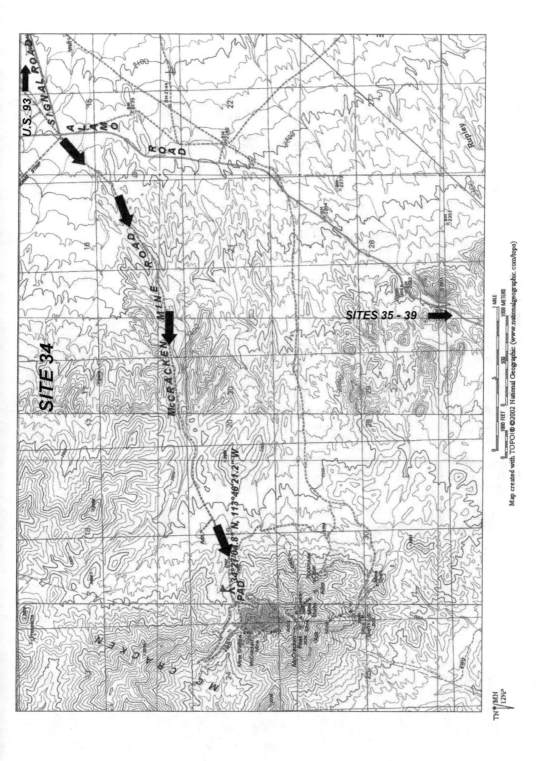

SITE 35

Chalcedony Lined Geodes near Keenan Camp

Difficulty Scale: 7 – 3 – 1   Seasons: Fall, Winter, Spring
Global Positioning System Coordinates: 34° 22' 59" N, 113° 46' 16" W*
Geology: Pliocene-Middle Miocene Volcanic Rhyolite and Andesite Flows
U. S. Geological Survey 7.5 Minute Topographical Map: McCraken Peak

FOLLOW THE DIRECTIONS TO SITE 34 on page 128 to the intersection of Signal and Alamo Crossing Roads. Turn left (south) at the intersection onto Alamo Crossing Road, drive 4.4 miles, and turn right (west) on the narrow road that follows the power line. After going just .1 mile, you will come to an intersection of three roads. Follow the middle road immediately to the left of the wilderness signs (Babbitt barriers) that goes up and over the little hill beside the power pole. Go west following the power line 1.1 miles to a fork. The left fork continues to

The four H shaped wooden pilings at the entrance to Keenan Camp.

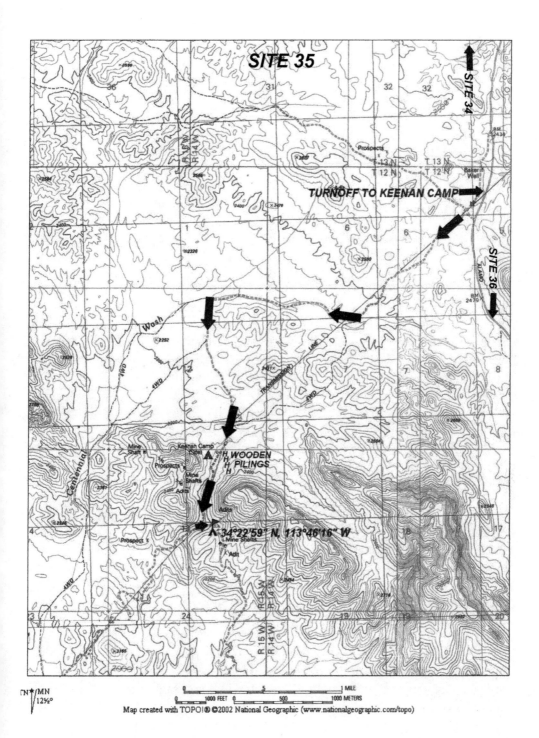

follow the power line, but becomes a steep roller coaster. Instead, take the right fork (north-west) into Centennial Wash and follow the road 2.1 miles around to the site of the old Keenan Camp. The road to Keenan Camp is pretty tame, but you may want to shift into 4-wheel drive in the sandy wash. When you get to Keenan Camp, you will see four H shaped sets of old wooden pilings lined up in a row. You will be on a flat at the foot of two small mountains. The collecting area is about .5 mile further on to the west. You will go through a gate, down a short steep slope, and into a rough wash that flows between the two small mountains. From here on, you will need four wheel drive. At the bottom of the wash, look for a steep, rough, narrow road to the left (south) that angels 200 – 300 yards up to the collecting area at the top of the hill. You will begin to see pieces of broken geodes in the wash. They will become more prevalent as you proceed up the narrow road to the top of the hill.

At the top of the hill, you will find the ground littered in all directions with geodes. The host rock here is a reddish-brown volcanic rhyolite that contains exposed geodes as well as chalcedony and crystal lined vugs. These geodes are typical of several other volcanic flows in several other Arizona locations. The material here, however, is more plentiful and of a generally higher quality than most other flows. The geodes have a roundish shape with a rough distorted exterior. The outside skin is an earthy muddy color similar to the surrounding country rock. They are visible in the igneous matrix. Some have weathered out far enough that you can remove them with a hammer and chisel. Often, you will find them clustered together like a clutch of eggs. Some are as small as marbles and others are as large as basketballs. When cut, the interior appears redder than the outside skin. Some of these geodes have solid quartz crystal centers and others open centers surrounded by chalcedony or quartz crystals. And, as with all geode patches, some are just solid rhyolite throughout—duds.

Collecting worthwhile geodes is as much a matter of luck as it is skill. Although, learning a few techniques can make your collecting efforts more successful. Broken geodes are, of course the easiest because you can see what is inside. You can select those that will trim and polish nicely. But, you will have only half a geode instead of a matched pair. If you desire a hollow center, select geodes that seem to weigh less than they should for their size. The quartz interior of these geodes is roughly star-shaped with one or more of the star points penetrating the outside skin. Look for geodes with evidence of chalcedony or quartz crystallization on the outside. Also, geodes that bear ridges dividing the geode sphere into thirds tend to have gemmy interiors. Avoid the temptation of smacking geodes with your rock pick. Even when you select geodes carefully in the field, chances are that only one in ten will be worth keeping after you cut it open. But, Murphy's Law says that if you randomly crack one open in the field, you will ruin a real keeper.

*G.P.S. coordinates taken at the collecting area.

SITE 36

Fluorite and Selenite at the Lead Pill Mine

Difficulty Scale: 7 – 4 – 6   Seasons: Fall, Winter, Spring

Global positioning System Coordinates: 34° 22' 36.8" N, 113° 42' 59.0" W*

Geology: Middle-Early Proterozoic Granitic Varieties and Biotite Granite

U.S. Geological Survey 7.5 Minute Topographical Map: Rawhide Wash

FOLLOW THE DIRECTIONS TO SITE 34, page 128, to the intersection of Signal and Alamo Crossing Roads. Turn left (south) on Alamo Crossing Road, drive 7.3 miles, and turn right (west) on the rough uphill road. This road is 2.8 miles south of the turnoff to Keenan Camp. See site 35 on page 132. Go .3 mile to a faint turnoff on your right (west) and park here. There is only enough room for one or two vehicles here. The road beyond this point is rough, narrow, and heavily eroded in several places requiring 4-wheel drive. From the parking place you will be able to see the Lead Pill Mine and the remnants of the road leading up to it on the mountainside to the southwest. Hike from the parking area in a southerly direction cross country to the low end of the road leading to the mine and follow it up to the Lead Pill from there. The distance to the road is about a quarter of a mile over moderately rough open terrain. Do not attempt to hike in a westerly direction directly toward the mine itself intersecting the road higher up the mountainside. The going is very steep and rough, the slope is thickly vegetated, and you will have to cross a deep arroyo. And do not attempt to drive up the mine road. It is narrow, deeply eroded, and there is no place to turn around.

Once you reach the end of the mine road, you will see the mine entrance ahead of and above you, a tailings slide in front of and below you, and a game trail up the slope toward the mine entrance in front of you. As you proceed up the game trail, you will quickly encounter evidence of fluorite and diggings. A fluorite bearing rock outcrop 10 – 20 feet wide runs up the mountainside at right angles to the trail. You must use a hammer and chisel to open the fluorite bearing seams. The matrix is fairly soft and fractures rather easily. The seams and vugs are lined with small (1/8 – 1/4 inch square) aqua colored fluorite cubes. If you pry the matrix apart carefully so as not to dislodge the fluorite crystals, you can harvest some decent specimens. Remember, to bring packing material to protect your specimens on the trip back.

The mine itself consists of a short trench that leads into a tunnel entrance. I do not recommend entering mine tunnels. They can be death traps. However, on the sides of the trench and in the tailings pile below are seams of white, chunky, opaque selenite. They look somewhat like compressed snow and ice that is decaying in a springtime thaw. If you are patient and careful, you can recover some interesting, crusty, crystally, half melted looking selenite sheets.

View of the route up to the Lead Pill Mine from the parking area.

*G.P.S. coordinates taken at parking area.

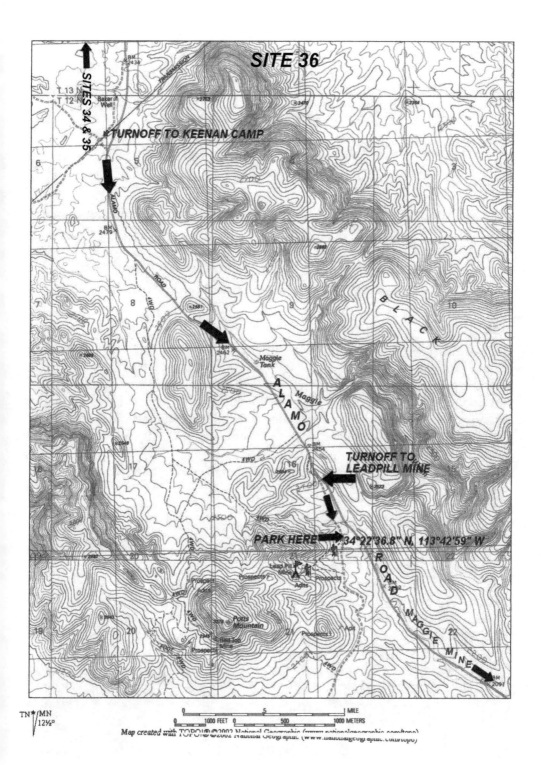

SITE 37

Malachite at the New England Mine

Difficulty Scale: 5 – 7 – 6   Seasons: Fall, Winter, Spring
Global Positioning System Coordinates: 34° 18' 50.1" N, 113° 42' 31.5" W*
Geology: Middle-Early Proterozoic Igneous Granitic Rocks
U.S. Geological 7.5 Minute Topographical Map: Rawhide Wash

FOLLOW THE DIRECTIONS TO THE MAGGIE MINE on page 336. But, instead of turning left (east) off Alamo Crossing Road, turn right (west). Go .75 mile to a fork. Before proceeding on to the New England Mine, you may want to take a little side trip here. Half way up the little canyon between the hills on your left (east) are some shallow digs containing veins of chrysocolla. This is not gem quality material, but you can pick up some nice colorful samples, yard rocks, or, if you feel energetic, you can attempt to bust out some better quality material from the heavily veined matrix.

To continue on to the New England Mine, take the right fork down into the

Looking across the ravine toward the adit.

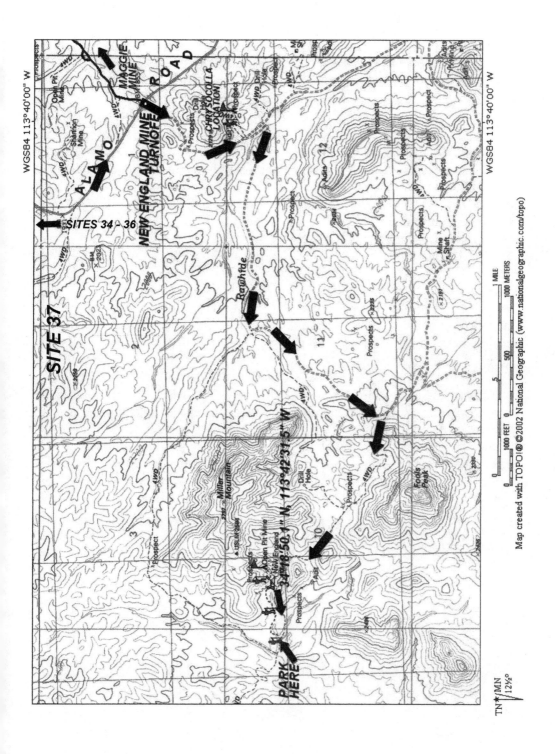

wash. Coming up the steep banking out of the wash on the other side may require four wheel drive. The rest of the trip is easily negotiated with two wheel drive high clearance vehicles. Proceed west 1.3 miles to another fork. Bear left (south) and continue on for another mile around the mountain to yet another fork. Bear right (west) at this fork and follow the road between the mountains. After going 1.4 miles, you will see an adit and small tailings pile on your right (north) across a deep little ravine that runs parallel to the road. Above the adit is a little rocky outcrop that runs up the mountainside at right angles to the ravine. On the left (west) side of the outcrop is a trench excavation that runs parallel to it. In front of the adit, the remains of a road go around the right (east) side of the outcrop and leads to the excavations at the top. This is your collecting destination.

The topographical map shows a road leading across the ravine to the adit and the diggings above. Alas, it is totally washed out. The washout is so deep here that you cannot even hike across the ravine. The best route to the mine begins a few hundred yards farther down the road to the west where the road crosses the ravine. From here, scramble up the north side of the ravine and hike back (east) to the adit. Be careful. Wear good sturdy hiking boots. The terrain here is steep, slippery, hard, and heavily vegetated with mean sticker bushes. Once you finally reach the adit, follow the road around the east side of the outcrop to the collecting area at the top. Here, you will find retaining walls, tailings piles, and excavations. Look through the rubble for manageable sized pieces that display layers of mossy looking malachite on one side. This malachite is not lapidary quality. It is not dense enough to work and its color is somewhat dull. But, it does make interesting mineral samples. The most attractive specimens have a layer of white quartz adjoining the malachite. Sometimes, the veins are porous and vuggy allowing the quartz and the malachite to intermingle.

The igneous matrix here is heavy, hard, and large. You will need a heavy hammer and a sturdy chisel to chop off as much heavy matrix from your specimens as possible. Be selective, remember, you will have to lug the stuff out on your back over the same terribly torturous terrain you traversed on the way in.

*G.P.S. coordinates taken at the mine.

SITE 39

Copper Minerals at the Cactus Queen Mine

Difficulty Scale: 5 – 5 – 4   Seasons: Fall, Winter, Spring
Global Positioning System Coordinates: 34° 17' 46" N, 113° 40' 22" W*
Geology: Middle-Early Proterozoic Granite Varieties and Biotite Granite
U.S. Geological Survey 7.5 Minute Topographical Map: Rawhide Wash

The narrow road leading up to the Cactus Queen Mine.

FROM THE INTERSECTION OF Signal and Alamo Crossing Roads (see Site 34, page 128) turn left (south) on Alamo Crossing Road, drive 15.5 miles passing the turnoffs to sites 35 – 38, and turn right (north-west). Follow this road 1.2 miles to a fork. On the mountainside on your left, you can see the extensive excavations of the Rawhide Mine. Turn left (west) at the fork, drive 1 mile, and turn left (south) again on the narrow road leading .25 mile to the Cactus Queen Mine.

With 4-wheel drive, you can continue on to the top of the hill where you can turn around. Otherwise, you can park and easily walk up the rest of the way.

Once you arrive at the top of the little hill where the mine shaft is, you will find ore dumps containing malachite and chrysocolla. In some of the dump rocks, these copper minerals form distinct veins that run through a dark, ruddy colored matrix. In others, the malachite and chrysocolla appear to be equally intermixed with the matrix creating a dark greenish or bluish material. The veined material makes good mineral specimens. The mixed material makes good lapidary rock.

The immensity of the Rawhide Mine workings that you saw on the way in to the Cactus Queen looks tempting. But, there is hardly anything worth collecting there. If you must have something from this location, there is a short, shallow trench at the end of the road that leads up to the mine. You will have to park at the base of the mountainside and walk a short way up the slope to reach it. The trench is the lowest excavation on the mountainside. Here, there is a little black, very vitreous calcite that fluoresces a very intense bright red.

G. P. S. coordinates taken at the parking area.

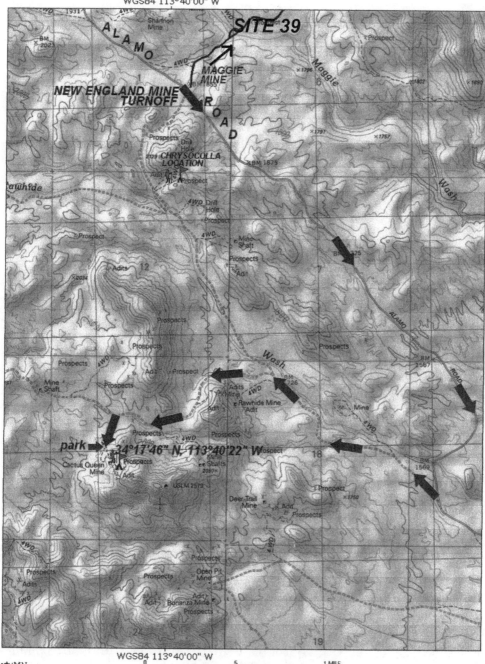

SITE 46

Obsidian on the Barry M. Goldwater Range — Former Area A

Difficulty Scale: 4 – 3 – 3   Seasons: Fall, Winter, Spring

Global Positioning System Coordinates: 32° 41' 06.4" N, 112° 18' 22.7" W*

Geology: Middle Miocene-Oligocene Volcanic Lava Flows and Tuffs

U. S. Geological Survey 7.5 Minute Topographical Map: Johnson Well

CERTAIN AIR FORCE AND MARINE CORPS areas of the Barry M. Goldwater Range (BMGR) are open to the public for recreational use. To gain access, you must first obtain a permit which authorizes you to travel to the Sand Tanks Area of the Sonoran Desert National Monument (formerly Area A of the BMGR), the Cabeza Prieta National Wildlife Refuge, the Sauceda Mountains Recreational User Area (Area B), the Ajo Air Force Station, The Bender Springs Area, and the western portions of the range not reserved for military live-fire and other training exercises. The application process takes about 30 minutes including a 22 minute film explaining the mission of the range and the rules you must follow when visiting it. Along with your permit, you will receive a lengthy pamphlet explaining the range does and don'ts, maps, instructions on how to properly use your permit, and the combination to the range gate locks. Permits are free of charge and issued for one year. Study your instruction pamphlet carefully and follow its directions precisely. To obtain a permit, you may call the Luke Air Force Base Gila Bend Auxiliary Field (928-683-6200), the Bureau of Land Management Phoenix Field Office (623-580-5500), the Arizona Public Lands Information Center (602-417-9300), the Yuma Marine Corps Air Station (928-269-2799), or the Cabeza Prieta National Wildlife Refuge (520-387-6483).

From the intersections of Interstates 8 and 10 in Eloy, go West on I-8 34.1 miles to Vekol Road (exit 144). Follow Vekol (Vekol Valley) Rd. south 8.3 miles and turn right (west) on the un-maintained dirt road opposite the large, abandoned, T-shaped building with the white roof on your left (east). After driving 4.15 miles on the dirt road, you will come to a fork in the road. Bear left (south-west) and continue on 1.9 miles to the eastern border of the BMGR. If

The parking area for location #1.

you fail to bear left, you will arrive at the wrong BMGR gate. As you approach the correct gate, you will see a corral and small shack on your left (south). Open the gate using the combination you were given with your range permit. From the gate, drive 1.8 miles where you will see two low hills beside the road on your left (east). This is the first obsidian deposit. Pull in and park beside the road (32° 39' 57.2" N, 112° 19' 27" W). The obsidian deposit is the crown of the hill farthest from the road. The second deposit is a huge obsidian reef jutting out from the base of the large mountain .7 mile farther down the road (32° 39' 49" N, 112° 21' 07" W). Park beside the road just before it crosses the large wash.

View of the obsidian reef from the parking area.

It is a short, easy hike along the east side of the wash to the western end of the reef. Here, you will find an obsidian rubble pile at the base of the reef containing pieces ranging in size from an inch to several feet across. The outside of most pieces is dull from weathering. The inside, however, is bright, glassy, and colored a deeply intensive jet-black. Although there are some imperfections and discolorations, there is a lot of massive, high quality, uniformly consistent material here. The "Area A" map that comes with your permit kit shows this area to be located immediately south of the Sonoran Desert National Monument in the unused portion of the BMGR under the jurisdiction of the Department of the Air Force. If you venture north, remember that collecting is not permitted in national park areas.

*G.P.S. coordinates taken at the BMGR entrance gate.

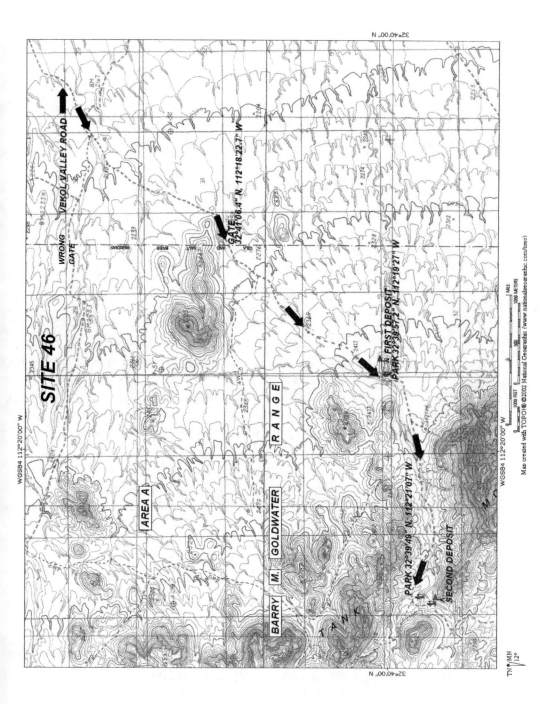

SITE 47

Chalcedony and Geodes on the Barry M. Goldwater Range — Area B

Difficulty Scale: 7 – 6 – 5   Seasons: Fall, Winter, Spring
Global Positioning System Coordinates: Given With Each Location
Geology: Middle Miocene-Oligocene Volcanic Lava, Tuff & Pyroclastic Rocks
U.S. Geological Survey 7.5 Minute Topographical Map: Hat Mountain

THIS SITE IS AN AREA COMPOSED OF FOUR separate locations in the Sauceda Mountains in Area B of the Barry M. Goldwater Range (BMGR) south of Gila Bend. The Sauceda Mountains are composed of thick felsic volcanic sequences, lava flows, and thick welded tuffs that form prominent, monolithic, physiological shapes. The four locations described below are found along area roadsides. If you are a hiker, you can probably discover many additional locations hidden back in the valleys and on the mountainsides. To gain access to the BMGR, see Site 46, page 144.

From the intersection of State Route 85 and the main street in Gila Bend, Follow S.R. 85 south 17.1 miles to Gate 6A which will be on your left (east). Open the gate using the combination you were given with your BMGR permit. Be sure to close and lock the gate behind you. Navigating the roads in the BMGR is easy. In your BMGR permit packet there is a map of Area B. Each intersection is numbered on the map and on a brown, flexible, plastic signpost in the field. From gate 6A, drive eastward 5.9 miles to intersection 3 and turn right (south). Drive 4.4 miles to intersection 5. Turn right and continue driving southward, past intersection 8, 3.1 miles to location A.

Location A. Chalcedony. G.P.S. Coordinates: 32° 37' 38" N, 112° 40' 59" W

As you approach each location, keep an eye out for traces of chalcedony along the roadside. Location A is the hillside on the left (east) side of the road. The chalcedony here is typical of other such deposits in Arizona. Look for exceptionally bright, shiny, drusey chips and unusually shaped pieces. Be

View of the Location A collecting area on the hillside east of the road.

prepared to hike uphill over a wide area. The higher up the hillside you go, the steeper and rougher the terrain becomes.

Location B. Chalcedony and agate. G.P.S. Coordinates: 32° 37' 22" N, 112° 41' 51" W

View of the slope leading up to the ridge at Location B. collecting area

Continue .4 mile southward from location A to intersection 11. Turn right (west) and drive 1.1 miles to location B which is the slope leading up to a low ridge on your right (north). The chalcedony here is much more varied than that at location A. As you ascend the slope, you will begin to encounter larger and larger plates and chunks of chalcedony. Some of these will have thick wavy agate rinds on the outside along with quartz crystal vugs on the inside. When you reach the ledge, look for agate and chalcedony veins coursing through the igneous host rock. Consulting your BMGR Area B map, you will see that this location can be approached from the west via intersections 19 and 13. Do not go this way unless you are a 4-wheel drive daredevil. There is a difficult mountain pass road between intersections 11 and 14.

Location C. Geodes. G.P.S. Coordinates: 32° 35' 59" N, 112° 39' 39" W

View from the road of the little cliff at location C.

When you have finished at location B, return to intersection 11. Continue past intersection 11 .9 mile to intersection 12 and turn right (south). After driving 1.9 miles, location C will be the slope and the little cliff above it on your right (west). This is the most interesting of the four locations. As you ascend the slope on the way up to the cliff you will cross a wash where you will begin to encounter geodes in the dirt. As you hike farther up, you will see that the cliff is a large obsidian dike chock full of rhyolite geodes 1 – 3 inches across. There is also a geode bearing rhyolite flow within the obsidian dike. The combination of earthy tan and ruddy colored geodes cemented together with shiny black obsidian is striking. Usually, the most desirable geodes are the ones containing crystally hollow centers rather than those that are solid rhyolite. Both types are

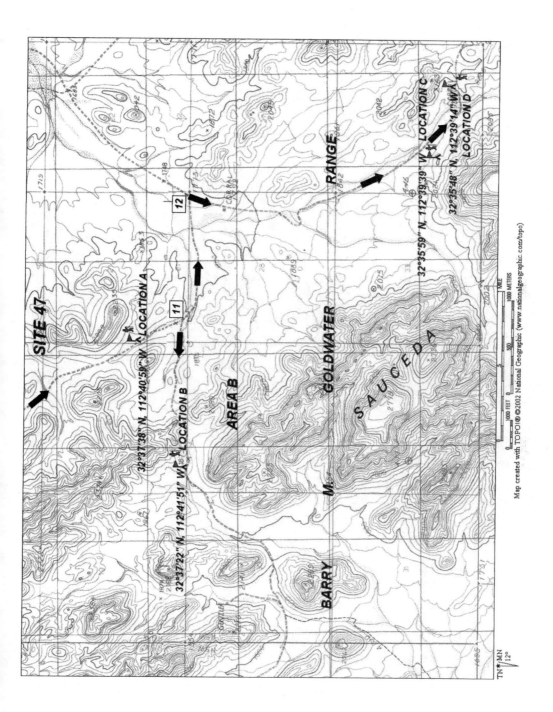

present here and both are equally desirable. The hollow ones contain good quality quartz crystals that are sharp and clear. The solid ones contain a core that displays unusual patterns and colors. Some have concentric circles like targets with a bull's eye in the center. Others have alternating rings of quartz crystal and rhyolite. There are some that have a pattern that radiates out from the center 360° like a star burst. Still others display a floral pattern.

Location D. Agate and Opal. G.P.S. Coordinates: 32° 35' 48.5" N, 112° 39' 14" W

The road leading up the hill beside Location D.

Location D is only .5 mile farther down the road from Location C. As you resume driving south, you will see the road ascending the side of a hill in front of you. Location D is at the top of the hill. Here, you will find some good quality wavy red and white agate. Look for large 2 – 10 inch long and up to 6 inch wide chunks. There are also good sized pieces in veins weathering out of the host rhyolite. The rhyolite flow sloping down hill from the north side of the road contains vugs with exceptionally sharp and clear quartz crystals. Also present here, in limited quantity, are veins of common white opal.

SITE 49

Chert at Battleground Ridge

Difficulty Scale: 3 – 4 – 6   Seasons: Spring, Summer, Fall

Global Positioning System Coordinates: 34° 29' 44" N, 111° 14' 27" W*

Geology: Permian Sedimentary Cherty Kaibab & Toroweap Limestone Formation

U.S. Geological Survey 7.5 Minute Topographical Map: Blue Ridge Reservoir

FROM THE TOWN OF STRAWBERRY on State Road 87 north of Payson, drive north on S.R. 87 10 miles to the Rim Road (Forest Road 300) also known as General Crook Trail. Running from the Verde Valley almost 150 miles across the Mogollon Rim, this road was built during the Apache Indian wars by the U.S. Army under the command of Brigadier General George Crook in the 1870's to move troops and supplies between Fort Whipple in the Verde Valley and Fort Apache in the White Mountains. Turn right (south) on the Rim Road and go 11.3 miles to F.R. 123. Turn left (north) on F.R. 123 which runs the length of Battleground Ridge and ends overlooking Blue Ridge Reservoir. Drive 3.5 miles to the beginning of the collecting area which is roughly a half mile long between the 3.5 and 4.0 mile points on F.R. 123. A power line crosses the road at the end of the collecting area.

Battleground Ridge is so named because it overlooks the site of the final battle of the Apache wars in Arizona. The battle occurred at Big Dry Wash, now part of Blue Ridge Reservoir, on July 17, 1882. A sizable band of Coyotero Apache Indians lead by Na-ti-o-tish left the White Mountains Reservation, killed Chief of Scouts J.L. Colvig at the San Carlos Reservation, attacked the mining camp of McMillenville, raided the Middleton Ranch, and killed a rancher named Bixby and his hired hand. Then, evading a trap set for them by volunteer rangers from Globe, they retreated through Tonto Basin into the forests of the Mogollon Rim. The U.S. Army pursued them with 14 troops numbering 1000 men lead by Captain A. R. Chaffee, an experienced Indian warfare veteran. Capt. Chaffee succeeded in surrounding his enemy on the rim of Big Dry Wash, a stretch of East Clear Creek, and engaged them in battle for about two hours. By the following day, 22 Indian fighters were found dead on the battle field. Many more were probably killed but never found. The survivors were captured and returned to the reservation. The U. S. force suffered 1 dead and several wounded. Lieutenant Thomas Cruse won the Medal of Honor for his bravery in this final battle.

When you reach the 3.5 mile point, you will begin to notice that the rocks in and along the sides of the road change from limestone to chert. Chert is not a remarkable or rare material and neither is its presence in the limestone formations of the Mogollon Rim. It is usually a pretty common, uninteresting, ordinary form of agate. The deep purple color of the chert at this location, however, makes it worth a second look as a lapidary material. As with most agate, it is dense, solid, cuts well, and takes an excellent polish. Some of the chert pieces are adorned with crystally quartz bubbles that look like warts. Occasionally, you may also find a brachiopod or two.

Collecting is easy here. All you have to do is get out of your vehicle and select material from the ditches beside the road. Or, you can collect in the woods on either side of the road. When you are finished here, you can proceed down the road to Site 48.

*G.P.S. coordinates taken at the beginning of the collecting area.

Looking down Forest Road 123 at the beginning of the collecting area.

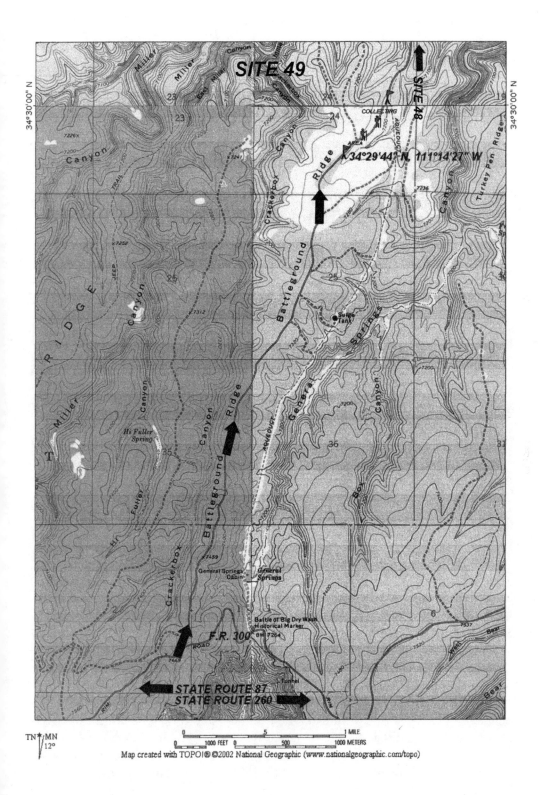

SITE 52

Pegmatites along the Black Pearl Road

Difficulty Scale: 3 – 6 – 5   Seasons: All
Global Positioning System Coordinates: 34° 41' 49" N, 113° 01' 29" W*
Geology: Middle-Early Proterozoic Granitic Varieties and Biotite Granite
U.S. Geological Survey 7.5 Minute Topographical Map: Behm Mesa

The beginning of the Black Pearl Road at the Bear Flat parking area.

FROM THE INTERSECTION OF STATE ROUTE 96 (Main Street) and North Lindahl in Baghdad, drive east on North Lindahl 1.6 miles to a fork. The left fork goes to the Phelps Dodge Hillside Mine entrance. Take the right fork, continue 2.3 miles up to the top of Nelson Mesa, and turn right (east) onto the road to Camp Wood. Before doing so, however, it is worth taking a few minutes for a scenic diversion.

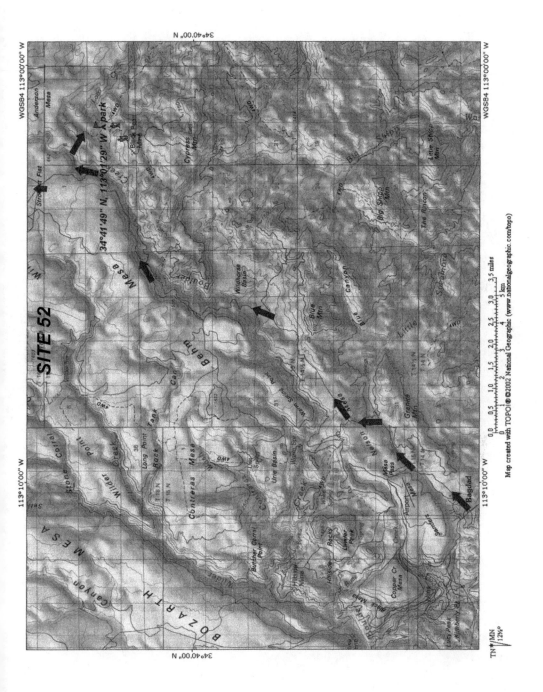

Go past the Camp Wood turnoff and continue on the rough road that leads north-west for .8 miles to a turnout where you can enjoy a spectacular view of Boulder Creek Canyon. Continuing your journey, drive east on the Camp Wood turnoff about 10 miles to the Black Pearl turnoff on your right. Look for the road sign. This road leads .9 mile down into the canyon where it crosses Boulder Creek. From here, continue following the road that leads straight ahead (south). Do not turn right onto the road that follows the creek westward. After driving 1.3 miles south, you will have to park at Bear Flat where the road begins to ascend the mountain. Beyond this point, the road up to the Black Pearl Mine is so badly eroded that even an ATV would have a difficult time. The hike from here is steep, long, and arduous and should only be attempted by those who are ready for a good cardio-vascular workout.

The mountains on the south side of boulder Creek are early Proterozoic metamorphic granitic rock containing abundant pegmatites. As you struggle up the steep washed out road, look for pegmatites exposed in the roadbed and in the eroded gullies in the surrounding woods. Most of these are just ordinary quartz and granite. But, there are some that are composed of a very aesthetic mixture of clear quartz, pink granite, and green epidote. Dense, gemmy, solid, and colorful, this rock makes an excellent medium for sculptors and lapidaries. If you make it all the way up to the Black Pearl Mine, you can collect milk-white quartz adorned with massive, silver colored iron pyrite.

G.P.S. coordinates taken at the parking area.

SITE 53

Malachite at the United States Mine

Difficulty Scale: 2 – 4 – 4   Seasons: Fall, Winter, Spring

Global Positioning System Coordinates: 34° 54' 47" N, 112° 15' 17" W*

Geology: Paleozoic Sedimentary Sandstone and Limestone

U.S. Geological Survey 7.5 Minute Topographical Map: Hell Point

View of the United Stares Mine from Forest Road 492A.

YOU CAN APPROACH THIS SITE FROM THREE DIRECTIONS. If you are coming from Jerome, travel west 14.2 miles on the Perkinsville Road (Forest Road 318) to the bridge across the Verde River at Perkinsville. Here, the road number changes to F.R. 354 and turns north. Continue northward on F.R. 354 for 5.6 miles and turn left (west) on F.R. 492. Follow F.R. 492 for 1.1 miles and turn left (south) on F.R. 492A. Follow F.R. 492A 2.8 miles to the United States Mine. If you are coming from Prescott, follow State Route 89 from the intersection of State Routes 69 and 89 33.7 miles and turn right (south-east) on F.R. 680 (County Road 71). If you are coming from Ash Fork, follow S.R. 89 from Interstate 40 17.1 miles and turn left (south-east) on F.R. 680. Drive 1.8 miles on F.R. 680 to Drake and cross the rail road tracks. From Drake, follow F.R. 492 10.3 miles to F.R. 492A, turn right (south), and drive south 2.8 miles to the mine.

There are hundreds of small, abandoned copper mines and prospects all over Arizona where you can find traces of malachite and other copper minerals. Very few of these, however, contain gem or specimen quality material. The United States Mine does not yield any gem quality malachite but, it does contain an unusual type of specimen. The strata between Ash Fork and Paulden are sedimentary Mississippian, Devonian, and Cambrian limestone, sandstone, shale, and mudstone. These pastel gray and cream colored strata tinted with subtle reds, yellows, and oranges, were formed by the intermittent rise and fall of early Paleozoic seas and are now mined in several large quarries for building and paving stone. If you can find an abandoned quarry, you may want to search among the 1 – 2 inch thick slabs for those displaying animal tracks. They are rare but you may get lucky. At the United States mine, there is a stratum of creamy yellow sandstone laced with deposits of massive, light green malachite. Sometimes the malachite forms encrustations across the face of the sandstone and sometimes it forms deposits reminiscent of random candle wax drippings. The combination of the textures and colors of these two materials together are pleasing and make an interesting addition to a specimen collection.

The United States mine is an open pit. Although collectable material is scattered throughout the property, the largest concentration is in the south-west part of the mine. Look in the roadways, bankings, gullies and rock piles for specimens. You can break up the larger rocks or dig out the smaller ones that are partially exposed in the dirt.

*G.P.S. coordinates taken at the mine.

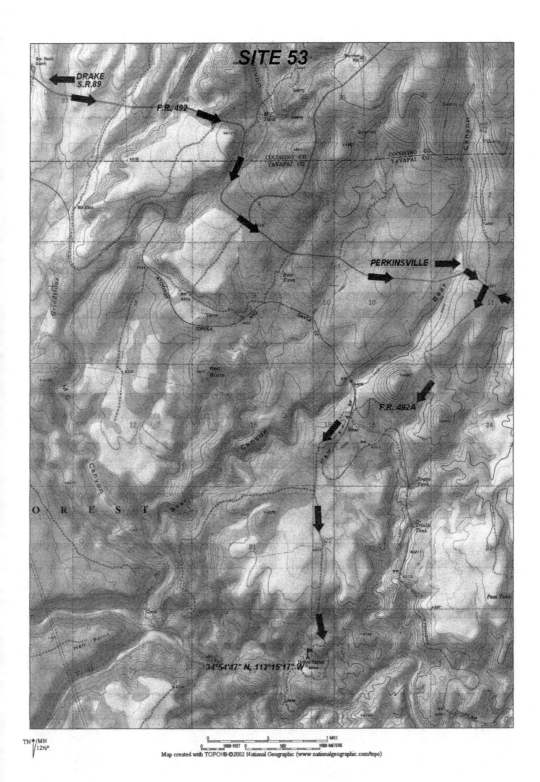

SITE 57

Marble off Ryan Ranch Road

Difficulty Scale: 5 – 3 – 2   Seasons: Fall, Winter, Spring

Global Positioning System Coordinates: 31° 51' 03" N, 111° 15' 54" W*

Geology: Jurassic-Cambrian Metamorphic Marble and Quartzite

U.S. Geological Survey 7.5 Minute Topographical Map: Penitas Hills

FROM THREE POINTS, ALSO CALLED ROBLES JUNCTION, take State Route 286 south 14.5 miles toward Sasabe and turn left (east) onto Ryan Ranch Road. Follow Ryan Ranch Rd. 7.7 miles to a fork at a cattle guard. Do not cross the cattle guard. Instead, bear left (east) on the rough, narrow road and follow it .6 mile to an even narrower trail that goes .2 mile up to the site. You can see the site about a mile or so before you reach the fork. It looks like a pile of white snow on the hillside ahead of you. Park beside the white pile.

View of the "pile of white snow" at the beginning of the narrow trail leading to the collecting area.

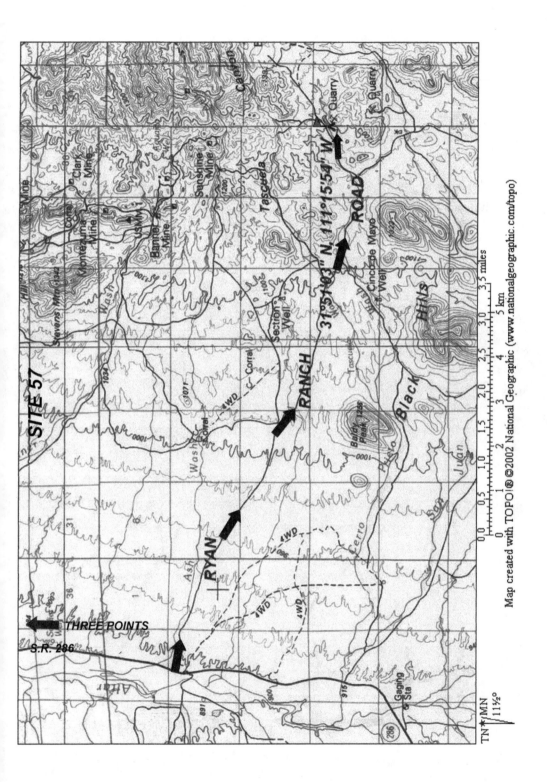

At the parking area, the ground is littered with small marble chips and pieces 2 – 3 inches across. The white pile is pulverized marble. Apparently, this was the area where large pieces of marble were loaded for shipment. The actual quarry where the stone was dug is on the north side of the hill. You can not see it from the parking place, but it is only a short walk up and around the hillside to reach it. There is no longer a road to the quarry. There are bigger chunks of marble there, but you are limited to what you can carry on your back. The quality of this marble is very good. It is solid, clear, bright, pure white, and free from impurities and stains. There is a second quarry about a mile farther south off Ryan Ranch Road, but it is about a mile hike to reach it. Nature has reclaimed the road up to it.

G.P.S. coordinates taken at parking area.

SITE 58

Geodes at Corral Nuevo

Difficulty Scale: 6 – 4 – 6   Seasons: Fall, Winter, Spring
Global Positioning System Coordinates: 31° 28' 55" N, 111° 10' 34" W*
Geology: Middle Miocene-Oligocene Volcanic and Pyroclastic Rhyolite Rock
U. S. Geological Survey 7.5 Minute Topographical Map: Ruby

FROM THE INTERSECTION OF INTERSTATES 10 and 19 in South Tucson, drive south toward Nogales on I-19 56.5 miles and turn right (west) on State Route 289. Follow S.R. 289 9.6 miles and turn left (south-west) on Forest Road 39. Follow F.R. 39 11.6 miles to F.R. 4186 where you will see a road sign saying "Corral Nuevo". Turn right (north) onto F.R. 4186 and follow it 2.3 miles to Corral Nuevo and park. To reach the collecting area from here, you will have to hike about .2 mile up the wash and into the canyon that runs eastward from the corral. The collecting area begins about 100 feet after the wash makes a sweeping 90° left turn to the north.

The entrance to the canyon as viewed from Corral Nuevo.

This may very well be the best rhyolite geode deposit in Arizona. The geodes here are large, contain excellent quartz crystal displays, and there are lots of them. The not so good news is that they are not very easy to harvest. This is evident when you arrive at the beginning of the collecting area. Here, on the bottom of the wash, you will encounter a layer of dark, solid rhyolite containing a lighter colored vein of material that is studded with geodes. These geodes, in the water course, have been severed in half by the powerful, debris laden stream flow. A little farther on, you will see a rhyolite dike that runs down the hill on the right side of the canyon, crosses the canyon, and runs up the opposite side. The best collecting is in the canyon wall on the right (east). Here, the geodes contain a variety of bright, delicate, shiny, quartz crystals. Thin needles, milky sprays, and gemmy-clear scepters sometimes form in the same geode. This dike makes a second appearance on the back side of the hill farther up the canyon where the wash makes its S curve back around to the south. On the opposite (west) side of the wash there is a dark-green obsidian dike. Unfortunately, it is not high quality material. It is course-grained and contaminated with earthy colored materials. As you continue to hike up the wash, you will encounter more geode bearing dikes and rubble and a little common agate that has washed downstream from higher elevations. When you arrive at the point where the canyon turns left (north-east) look for cantaloupe sized geodes in the cliff face on your left and in the large boulders lying in the stream bed. Some of these geodes contain zones of common, opaque, white opal.

Bring your sledge hammer and chisels with you. Since most of the geodes here are still firmly encased in their tough rhyolite matrix, you will have to exert some extra energy to harvest them. Geodes of all sizes are highly concentrated and tightly cemented together by the rhyolite flows. Look for small boulders and rocks in the stream bed that nature and previous rockhounds have broken loose from the dikes. Even if you can not separate geodes from this material, you can trim the rock down to smaller pieces displaying 2 or 3 nice geode halves. A lot of geodes in the dikes in the canyon walls are half exposed. If you are careful, you can liberate some of these intact with your rock pick or with a hammer and chisel. Once free from their matrix, these geodes can then easily be cut and polished.

G.P.S. coordinates taken at Corral Nuevo.

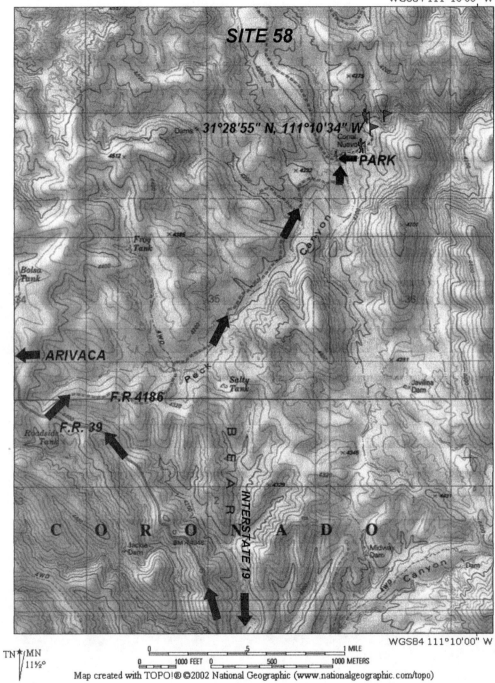

SITE 60

Chrysotile at the Phillips Mine

Difficulty Scale: 3 – 5 – 3   Seasons: all

Global Positioning System Coordinates: Given with each location

Geology: Middle Proterozoic Sedimentary Grand Canyon & Apache Groups

U.S. Geological Survey 7.5 Minute Topographical Map: Mule Hoof Bend

FROM THE INTERSECTION OF U.S. Route 70 and State Route 77/U.S.60 in Globe, go north 33.1 miles on S.R.70/U.S.60 and turn left (west) at the old Seneca Lake trading post. Drive .8 miles past Seneca Lake and over the spillway to the intersection of F.R. 1302. The pavement turns right (north) to an overlook of the Salt River Canyon. Proceed straight ahead west on the unpaved F.R. 1302, which soon becomes F.R. 473, another 2.95 miles and you will come to the little Phillips Mine ghost town. Turn in and park anywhere you like. There is no one there to complain.

The turn in to the Phillips Mine ghost town from Forest Road 473.

The Phillips Mine G.P.S coordinates:
33° 47' 21" N, 110° 32' 48" W*

The collecting area is the tailings slide on the small mountain on the northeast edge of town. Mother Nature is desperately trying to reclaim the short road from town to the mine, but you can still drive to the base of the mountain if you really want to. Otherwise, it is a short walk to the bottom of the tailings slide where the best material is found. The slides above and below the road that leads up to the second level are steep and loose. If you follow the road up to the second level at the top of the slide, you will come to a wide bench with several adits penetrating the mountainside. In addition to serpentine, there are small occurrences of calcite and aragonite here as well.

The collectable material here is green and white serpentine laced with thin vitreous veins of chrysotile commercially known and mined as asbestos. The geology and chemistry of this kaolinite-serpentine mineral group is very complex and still the subject of continued research. Serpentine is a secondary mineral typically consisting of polymorph mixtures derived from the alteration of magnesium silicates. It is formed in limestone and hydrothermal veins. See pages 20 and 178 for more on this subject.

This area north of Globe along the Salt River Valley was the most commercially important asbestos mining area in Arizona. There are several other mines and prospects in the vicinity of Seneca Lake to explore. The Phillips Mine is the easiest to reach. The Emsco and the Regal Mines are also accessible from F.R. 473. The Canadian Mine just north-west of the Phillips is posted with a private property sign.

G.P.S. coordinates taken at the turnoff to ghost town.

The Verde Antique Marble Area G.P.S. coordinates:
33° 47' 11" N, 110° 36' 35" W**

West of the Phillips Mine, you can collect what appears to be verde antique marble-metamorphosed serpentine. From the entrance to the Phillips Mine, Continue west 2.5 miles on F.R. 473 and turn left (south) on F.R. 473A. This is a rougher road than F.R. 473 requiring high clearance. The collecting area begins 1.8 miles down F.R. 473A. The area is fairly flat and littered with large rocks and ledges. Because of weathering, these rocks are dull, grayish-brown in color on the outside, and generally uninteresting to look at. You will need a sledge to break these rocks open and reveal the true quality within. This rock is dense, free from vugs, and uniformly textured. It is more streaked than striped with equal portions of subtle light green and grayish-white tones. It is very handsome and would be of interest to carvers and sculptors.

The beginning of the verde antique collecting area on Forest Road 473A.

**G.P.S. coordinates taken at the beginning of the collecting area.

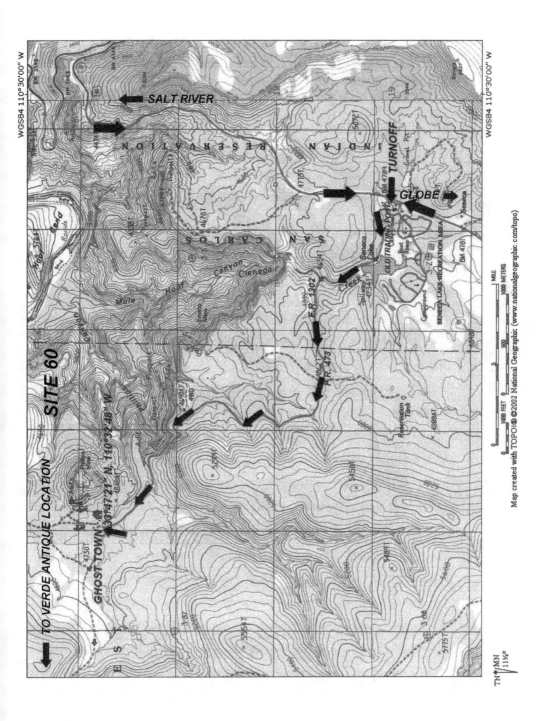

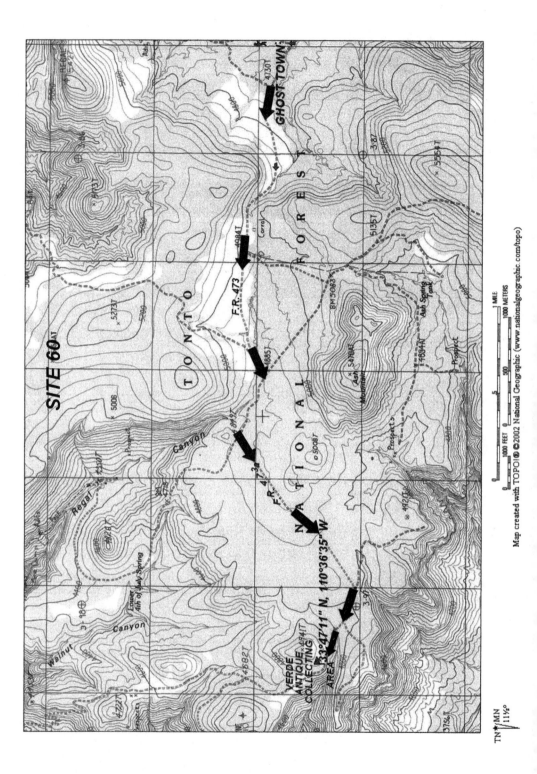

SITE 61

Onyx on Forest Road 303

Difficulty Scale: 2 – 3 – 3   Seasons: Fall, Winter, Spring
Global Positioning System Coordinates: 33° 41' 30.3" N, West 110° 39' 20" W*
Geology: Jurassic Igneous Intrusive Granitic Rock and Quartz Diorite
U.S. Geological Survey 7.5 Minute Topographical Map: Haystack Butte

FROM THE INTERSECTION of U.S. Route 70 and State Route 77/ U.S.60 in Globe, drive northward on S.R.77/U.S.60 16.7 miles and turn left (west) on F.R. 303 (also known as Haystack Butte Road). The turn is at the end of a guard rail across the road from the Jones Water Campground. F.R. 303 is a well maintained gravel road suitable for passenger cars. After driving 10.2 miles northwest on F.R. 303, you will come to a small parking turnout on your right beside a rough, narrow rut that used to be the road leading up to the onyx bearing ledge on the ridge above you. Park here and walk up to the collecting area.

Parking area and rough, narrow rut leading to collecting area.

This site has been well-known within the rockhound community for years. A collecting book published as early as 1956 makes reference to it. It was not included in *Minerals of Arizona* because when it was published, there was hardly anything accessible to collect there. Near the top of the ridge, a ledge of pure onyx several feet thick was visible but it was so solid and resilient that it was nearly impossible to harvest anything from it. All that has changed now. The onyx ledge is still there, but someone has recently drilled and blasted it releasing a good amount of onyx rubble for collectors to rummage through. There are even saw marks on some of the large boulders that were blasted loose from the onyx seam. So, we now have a reborn collecting site.

The onyx here is excellent for lapidary and sculpture. For the most part, it is thick, dense, and free from cracks and voids. There is some, however, that contains bright, white crystally vugs. The stone is multi colored displaying several shades of red, salmon, orange, yellow, brown, white, and black. Strata are arranged in parallel, wavy, random, and intersecting patterns. When cut or split, these various patterns and colors can produce some very unexpected and pleasing visual effects.

Working here is not terribly difficult unless you attempt to break up some of the larger boulders. Hiking up the short washed out road to the collecting area is tedious. But, once you have reached the top, the horizontal road that runs the length of the onyx vein is flat and easy. There is considerable material on this road that can be easily collected. Some larger pieces, however, have tumbled down the slope below the road. Retrieving these on the steep brushy banking is more difficult. When you reach the top, you can go left or right on the horizontal road. To the left, you will find the more colorful onyx in the area that was recently blasted. To the right, you will find a more subdued white, tan, and cream colored stone.

Whatever variety you choose, consider taking only the amount you can reasonably expect to use and leave the rest for those who will follow. The available supply is not inexhaustible. This site did dry up once before. If people get greedy, it can do so again.

*G.P.S. coordinates taken at parking turnout.

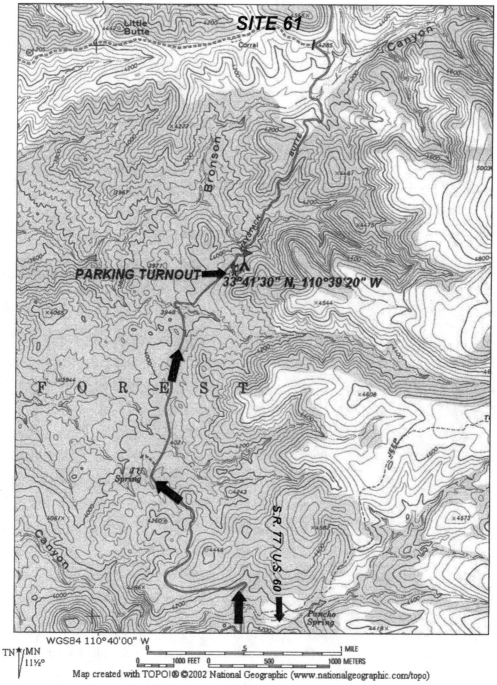

SITE 62

Serpentine on Forest Road 189

Difficulty Scale: 4 – 5 – 6   Seasons: Fall, Winter, Spring
Global Positioning System Coordinates: 33° 43' 52.6" N, 110° 56' 10.3" W*
Geology: Middle Proterozoic Sedimentary Grand Canyon & Apache Groups
U.S. Geological Survey 7.5 Minute Topographical Map: Meddler Wash

The turnoff to the parking area from Forest Road 189.

FROM THE SALT RIVER BRIDGE on State Route 288 at the east end of Theodore Roosevelt Lake, drive north 11.6 miles and turn right (east) onto Forest Road 189. Follow F.R. 189 2.1 miles across Worm Creek and up and around the south and east side of a steep hill where you will come to a little narrow road on your right. This unmarked road leads down onto a flat area where you can park and turn around. The road to the collecting area is about 100 feet farther up F.R. 189

on the left (west) side of the road. It curves horizontally back around the south and west side of the hill at about the 4200 – 4300 foot level overlooking the road that you drove up. Do not try to drive anything bigger than an all terrain vehicle down this road. Erosion and overgrowth has reduced it to a foot path and there is no place to turn around. F.R. 189 is in good condition but beyond Worm Creek, it gets steep and narrow. The next turnaround beyond the trail to the collecting area is about .5 mile.

This collecting area is one of the many asbestos prospects common to the Sierra Ancha mining district north of Lake Roosevelt. In fact, as you proceed north on S.R. 288, you will clearly see two long tailing slides resembling rabbit ears near the top of Asbestos Peak in front of you. The collecting hill is composed of chalky, white Pliocene to middle Miocene limestone laced with dark and yellow-green serpentine and horizons of silvery and brassy chrysotile. As you begin to hike the trail to the collecting area, observe the limestone rubble on the trail and the strata above it. Generally, it is chalky and unremarkable. Some pieces, however, appear harder and denser and display subtle patterns of swirls and stripes similar to rhyolite. When the trail turns the corner and takes you northward, you will begin to encounter better quality material. Rocks are larger, more solid, denser, and more uniform in texture throughout. Pleasing pastel colors have been added to the swirls and stripes resulting in exceptionally attractive displays. You will also begin to discover large veins and intrusions of serpentine coursing through the harder limestone. The serpentine has a resinous luster and a rich, deep color. And finally, as you approach the mine excavations at the end of the trail, you will come upon large limestone rocks containing bright, silky chrysotile horizons. These are often accompanied by alternating layers of gemmy serpentine. The chrysotile fibers are usually arranged in a tightly packed, perpendicular, wheat-sheaf fashion. However, some are wavy or bent and rest at a 30 – 45% angle to the horizon. Near the end of the trail, you will find a tall adit cut into the side of the hill. The entrance and roof of this tunnel are severely fractured and look as if they are going to cave in at any minute. Do not go in. It is just too risky. Below you are tailings piles, roads, benches, and another small tunnel supported by timbers that look as solid as stale bread. These areas offer ample collecting opportunities. These were asbestos and uranium mines bearing the names: Black Diamond, Clark Property, Friday Claim, Globe Group, Miami Group, and Rainbow Deposit.

Collecting here is challenging. Specimen collectors will have the easiest time of it. Small, cabinet size pieces are readily available on the road and in the dumps. These can easily be collected and carried away. Those who desire larger pieces for sculpture, lapidary, or slabbing, will have a more difficult job. You will need to do some sledge hammer work to reduce the larger boulders to pieces small enough to haul out on your back. A lot of good material has tumbled down the slope below the trail making it very difficult to retrieve. The slope above the trail contains some excellent serpentine and chrysotile bearing strata. However, the slope is steep and the ledges are thick and tough making extraction difficult and laborious. Depending upon how ambitious you are, this site can easily become a 7 – 10 on the site difficulty scale.

Serpentine is a very complex mineral, or perhaps set of minerals, whose exact composition is still under study. It is a magnesium, iron, or aluminum hydroxide composed of complicated polymorphs such as parachrysotile, clinochrysotile, lizardite, and others. Serpentine results from the alteration of such minerals as olivine, pyroxene, and amphibole. It is usually formed through contact metamorphism in magnesium limestone. The most common serpentine found in Arizona is chrysotile which was mined in large quantities for asbestos.

G.P.S. coordinates taken from the intersection of Forest Road 189 and the trail to the collecting area.

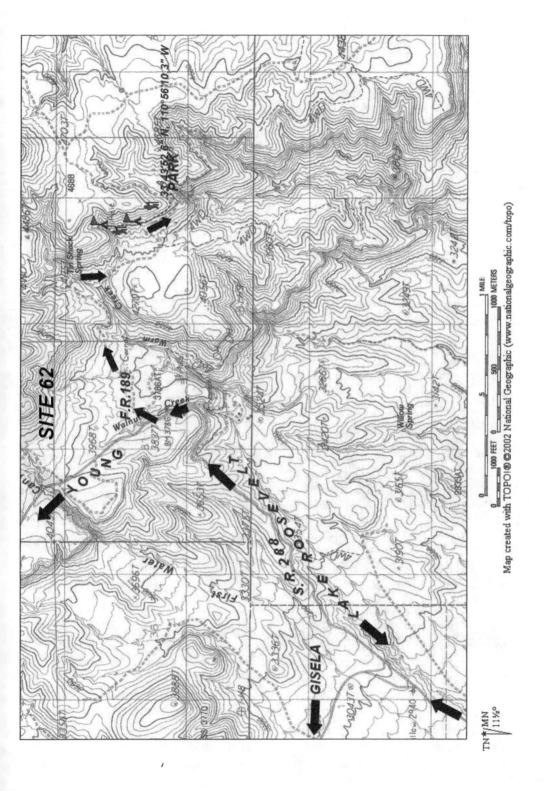

SITE 64

Amethyst at the Woodpecker Mine

Difficulty Scale: 7 – 5 – 6   Seasons: Fall, Winter, Spring

Global Positioning System Coordinates: 33° 12' 34" N, 111° 11' 40" W*

Geology: Early Proterozoic Metamorphosed Yavapai Supergroup and Pinal Schist

U.S. Geological Survey 7.5 Minute Topographical Map: Mineral Mountain

View of the Woodpecker Mine collecting area from the top of the ridge.

YOU CAN APPROACH THIS SITE FROM two directions. Whichever way you come, the trip is long and slow over rough roads. You must pay close attention to the directions because at certain points, there are several roads, not all of them shown on the topographical map, leading off in all directions. If you are coming from Florence via S.R. 79, follow the directions to Site 63 on page 366. Continue

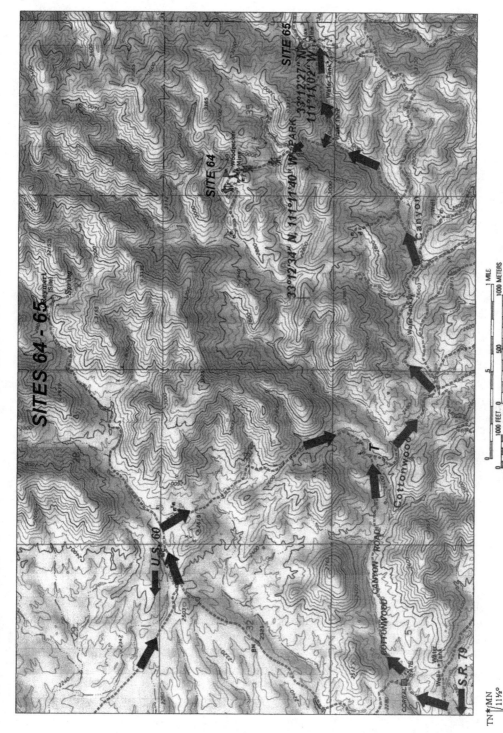

eastward past the turnoff to site 63 .7 mile to a gate at a large corral. Road maintenance ends here. From the gate, go 1.4 miles to a T. If you are coming from U.S. 60, travel east 3.6 miles from Florence Junction and turn right (south) on the gravel road. After driving 4.4 miles, turn left (west), go .3 mile, and turn right (south). Follow this road 1.5 miles to the T. The G.P.S. coordinates at the T are 33° 12' 10" N, 111° 13' 28" W. From the T, travel south-eastward .7 mile to an intersection of several roads. You can see them snaking over the ridges to the east. At this juncture, you will be going in a northerly direction. Do not make any turns. Continue strait ahead. After going .6 mile, you will come to a windmill and water tank in a wash. The turnoff to the Woodpecker Mine is .8 mile beyond the windmill. The road to the Woodpecker Mine will be a sharp left turn up a steep hill. You will need 4-wheel drive to make it up the rough, washed out road to the top of the ridge where you can turn around and park. The distance to the top is .3 mile. From here, you can survey the many Woodpecker Mine tailings piles, trenches, and old roads spread out over the hills and valleys to the north. If you are a determined 4-wheel drive aficionado, you can continue to drive down the severely eroded road before you that leads down the hill to the large tailings pile in the gully below. If you are not so inclined, you can hike down.

    The best collecting area begins at the tailings pile in the gully and continues up the road that goes past it to the mine shaft at the top of the hill. Look for dark, earthy colored, massive rock laced with more or less clear, crystally quartz veins along the roadside and in the wash on the west side of the road. This rock comes from an exposed reef on the hill on the east side of the road. What appears to be just ordinary quartz on the outside is actually beautiful, blue, massive amethyst on the inside. In the intense Arizona sun, the exposed amethyst surfaces faded.

    You can attempt to break up the large boulders with a sledge hammer or bust the smaller rocks with a 3 – 5 pound hammer. Much of this rock is thick, solid amethyst that makes excellent slabbing material. Therefore, you may want to test rocks you suspect to contain amethyst by carefully chipping off a small corner with your rock pick to see what is inside. This way, you will not fracture good quality lapidary material.

*G.P.S. coordinates taken at the top of the ridge.

SITE 68

Vanadinite at the Grey Horse Mine

Difficulty Scale: 7 – 5 – 3   Seasons: Fall, Winter, Spring

Global Positioning System Coordinates: 33° 05' 55" N, 110° 56' 01" W*

Geology: Pliocene-Middle Miocene Sedimentary Sandstone, Limestone, Mudstone

U.S. Geological Survey 7.5 Minute Topographical Map: Kearny

From the intersection of U.S. Route 60 and State Route 177 in Superior, travel 18.6 miles south on S.R. 177 where you will come to Ray Junction Road on your right (west). The turnoff to the Grey Horse Mine is on your left (east) about 50 feet beyond Ray Junction Road. It is an unmarked dirt road that leads uphill across a cattle guard. This turnoff is easy to miss because it is at the top of a rise and invisible until you are almost abreast of it. The distance to the mine is 2.4 miles. The road is narrow, well-drained, and may require 4-wheel drive for a short distance. There is room to park and turn around above the wash opposite the mine entrance.

This site has something for everyone; minerals, fossils, and fluorescents. About 1.5 miles up the road to the Grey Horse Mine, you will begin to see rounded battleship grey colored Pliocene-middle Miocene epoch fossiliferous limestone rocks scattered among the sandstone, shale, mudstone, and limestone deposits that are eroding down the west slope of the Dripping Spring Mountains. These fossiliferous rocks are widely distributed over the landscape on both sides of the road all the way up to the mine and beyond. Scattered across the surface of some of these rocks is marine fossil hash containing primarily brachiopod and crinoid parts as well as occasional mollusks and sea urchin spines. Petrified coral has also been reported in this region. You can collect fossils close to the road or, if you feel energetic, you can explore the steep slopes and deep ravines in the surrounding area.

There are several minerals available for collecting at the Grey Horse Mine, vanadinite being the most prominent. You can easily collect calcite and massive hematite. In the past, cerussite, descloizite, and wulfenite have also been discovered here. Search the tailings for veins containing calcite combined with vanadinite. The vanadinite crystals are small and occur as drusey coatings on the surface of the mine rubble. The vanadinite here is only semi-lustrous, but what it lacks in brilliance, it makes up in color. Varieties include bright red, yellow, orange, brassy, light brown, and blackish brown. Frequently, vanadinite druses will be overlain with a thin layer of colorless translucent calcite allowing the red and orange vanadinite color to show through. Sometimes, small individual vanadinite crystals will appear encased in or perched on top of the calcite. The host rock is laced with calcite-vanadinite seams. When you strike this rock with your rock pick, it tends to split along these seams revealing fresh, clean specimens. Along the wash below the mine and on the steep slope above it, look for fossils and chunks of massive hematite.

Although the calcite seams at this location will fluoresce, they are not spectacular. The edges of the calcite overlays tend to fluoresce a light pink color. The color is pleasing, but not intense or particularly bright. On the other hand, there is also a lesser amount of very brilliant blue-white fluorescing scheelite among the tailings.

*G.P.S. coordinates taken at the turnoff to the Grey Horse Mine from S.R. 177.

The approach to the Grey Horse Mine.

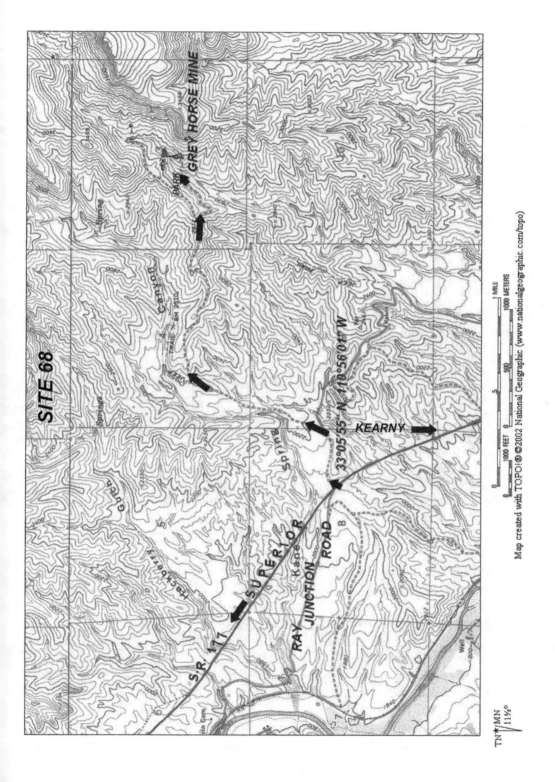

SITE 70

The Copper Creek District

Difficulty Scale: 7 – 8 – 7   Seasons: Fall, Winter, Spring
Global Positioning System Coordinates: Given with each location
Geology: Early Tertiary-Late Cretaceous Igneous Granitic
Diorite and Porphyry
U.S. Geological Survey 7.5 Minute Topographical Maps:
Oak Grove Canyon and Rhodes Peak

THE COPPER CREEK DISTRICT is an interesting and adventurous destination. However, because it is remote, rugged, and primitive, only experienced outdoors people and skilled 4-wheel mountain road drivers should attempt this trip. Even experienced wilderness adventurers should not go alone and traveling with more than one vehicle is advisable. Copper creek and Bunker Hill Roads are occasionally used by cattlemen and 4-wheel drive hobbyists, but the roads leading off into the side canyons and mountains are seldom traveled. These roads are covered with weeds, littered with fallen rocks and stones, and are usually uncomfortably free of tire tracks. Some roads here are well on their way to being reclaimed by Mother Nature. Despite years of neglect, however, these roads are, for the most part, in surprisingly good shape. Four-wheel drive and high-clearance is required but only on certain steep grades and on a few rough areas. As of the publication date of this book, all of the roads to the three collecting locations were passable. However, natural forces may change that in the future. Whenever you are uncertain, hike up the road ahead to be sure you can negotiate the way safely. Do not get trapped by a landslide or a washout with no room to turn around. Trying to back down a steep, narrow mountain road is no day at the beach. Be especially suspicious of the many old, obscure, overgrown roads that do not appear on the map and lead off in all directions to unknown destinations. Many of these lead to nowhere and stop dead half way up a steep mountainside.

Copper mining began here in the 1880's. By 1907, the town of Copper Creek had grown large enough to warrant a post office which remained open until 1942. In 1910, the town had a population of about 200, more than fifty buildings, its own stage line, and three operating copper companies: Calumet & Arizona, Copper Creek, and Minnesota Arizona Mining. Today, only a few concrete foundations remain.

View from the road leading out of the wash and up
the mountain to Location # 1

Hiking and climbing in this area also requires skill and good physical endurance. The creek beds and washes are generally deep ravines filled with large boulders and choked with thick stands of pricker bushes. The mountain and hillsides are equally difficult with steep slopes, slippery ground, thick vegetation, and large obstacles blocking the way. Mine excavations are very treacherous. Tailings piles and slides are as steep, tall, and unstable as any in Arizona. The slide surfaces are too hard to get a foothold in and dislodging one rock could cause a serious landslide. The tailings can not be safely explored.

To reach the Copper Creek district, first go to the corner of Main and Bluebird Streets in the tiny town of Mammoth on State Route 77. Go east on Bluebird St. across the San Pedro River until you come to a stop sign at the intersection of Bluebird St. and River Road. Copper Creek Road is the gravel road leading eastward on the opposite side of River Road. Follow it eastward for 9 miles until you come to a hairpin turn at the end of a long downgrade. This is the entrance to the Copper Creek Canyon. Continue into the canyon .95 mile where you will come to a road on your left (north) leading down to and crossing Copper Creek.

This road leads to locations 1 and 2. Just ahead and on your right (south) will be two toxic looking holding ponds surrounded by a chain link fence. The turnoff to location 3 is 2.3 miles farther up Copper Creek Road from the turnoff to locations 1 and 2.

Location #1 G.P.S. Coordinates:
32° 45' 10.4" N, 110° 29' 05.7" W*

To get to location 1, turn left off Copper Creek Road onto the road that leads north-eastward down and across Copper Creek. Follow this road uphill around the bottom of the huge Reliable Mine tailings slide above and a deep wash below. It is said that the blast that caused this slide is the largest non-nuclear explosion ever detonated. You will need 4-wheel drive on this stretch of road. After going .45 mile, you will come to a gate beside a dam that crosses the wash. On the opposite side of the dam is a washed out road with a chain link gate across it. Just as well, because it is one of those roads that leads to oblivion. Continue northward on through the gate another .15 mile through the wash to the point where the road turns westward uphill out of the wash. Continue strait ahead (north) across the wash on the road that leads to the Copper Prince Mine. After driving uphill about 100 yards, bear right onto the road that follows the draw upward in an easterly direction. Location 1 is the mountaintop visible above you. You can see the road you are on angling up the west side of the mountain. Follow it up and around to the very top. This road is not shown on the 7.5 minute topo map, but the map on page 191 has been edited to include it. The total distance from the turnoff in the wash and the top is .9 mile.

About 100 yards before reaching the summit, you will see a road to the right leading downhill in an easterly direction. This is the road to Location 2. Continue upward to the mountaintop where you will see a rocky knob on your left (west) and a flat area on your right (east). The flat area makes an excellent campground. It provides a grand view of the entire Copper Creek district. From here you can get your bearings and see how the roads wind through the district. Directly to the west, you can see the terraced Reliable Mine excavations covering the entire mountainside from top to bottom. But, no need to go there. Despite the enormity of this mine, there is nothing worth collecting there. After blasting and excavating the underground workings, the mountainside was terraced into leach fields. The leaching process dissolved the copper minerals leaving behind nothing but dirty yellowish rubble. Below the road, you may find some copper minerals in the rocks that tumbled down into the wash and escaped the leaching.

The location 1 collecting area is the south-east side of the rocky knob and the slope below it. This area yields a variety of quartz crystal vugs and seams in a resilient granitic matrix. Most of the crystals are cloudy or slightly milky with all six sides roughly equal in width. There are Japanese twin tabular crystals available here. They are rare, somewhat clearer than the more ordinary crystals, and may be small. Search carefully among the seams and vugs and you may be lucky enough to find one or two lurking among the more common crystal clusters. There are also some very attractive malachite stained and encrusted quartz crystal vugs. In addition to the secondary malachite washes, some vugs are also gilded with bright pink lepidolite deposits. As if that were not enough, many specimens are additionally adorned with needles and sprays of blackish-green tourmaline crystals. Some of these are even encased in the quartz crystals.

The terrain at this location is a 6 on the Difficulty Scale. It is slippery, heavily vegetated, steep in some places, and obstructed by large cliffs and boulders. You can choose how hard you want to work. You can try to liberate crystals from the huge boulders and cliffs with chisels and sledges, an 8 difficulty rating. Or, you can search through the considerable tailings left behind by previous rock hounds and the smaller pieces that have naturally eroded onto the slope below—difficulty 3.

Location #2 G.P.S. Coordinates:
32° 45' 10" N, 110° 28' 55" W*

From location 1, start back down the road you came up but instead of going westward back down the mountain to the wash, follow the road that leads downward to the east. Within about .2 mile, you will come to a wide flat area. To this point, the road is about a 6 on the Difficulty Scale. From this point on it is about an 8. You will have to decide to either park and hike down or 4-wheel it the rest of the way. There is a turnaround at the bottom. You will not be able to see the collecting location until you are about half way down the road. Then, two awesome quarries will come into view. They look like two gigantic bomb craters blown out of the mountainside. The road ends at the entrance to the first quarry. You can walk right in. The second quarry to the south, however, is inaccessible. In the first quarry, the collectable material is on the east and north sides. The west and south sides are nothing but large piles of broken country rock. As you enter the quarry, turn right. Walking northward, you will first encounter ruddy matrix containing blotches of shiny, silvery galena. Farther on, you will come to a slide of lighter colored material containing vugs of calcite crystal that have been stained and encrusted with secondary malachite deposits. The combination of the two minerals is quite attractive. The calcite crystals come in at least two habits; dog-tooth and tabular. As an added bonus, the calcitic matrix fluoresces bright red. You will have to dig into and rake the slide to uncover specimens.

Location #3, The Green Rock, G.P.S. Coordinates:
32° 44' 49" N, 110° 28' 38"W*

To reach location 3, continue on Copper Creek Road from the turnoff to location 1 2.3 miles through the deepest part of Copper Creek Canyon to a fork in the road. You will climb two rough 4-wheel drive hills along the way. One is before and one is after the Childs-Altwilkle Mill ruins. At the fork, there will be a concrete wall between the two roads bearing the name "Copper Creek". Take the road uphill to the right. Go .45 mile and park in the flat area beside the large reddish malachite stained promontory. From a distance, this huge outcrop looks as if someone had painted green blotches and stripes over the rock and cliff faces. Closer inspection reveals malachite deposits, apparently along old water courses, throughout the rock's cracks and crevices. These deposits are so numerous, that when viewed from the ridge above to the south, the rock actually looks like a green giant.

Specimen recovery here is somewhat difficult because the rock matrix is hard and the malachite deposits are soft, thin, and tend to be crumbly. At the base of the rock on the south side is a short adit. The best method is to find a spot where you can get a flat chisel or lever into a crack and pry off 2 or 3 inch malachite sheets. You may have to stabilize the matrix backing your specimens to prevent them from crumbling away.

If you continue 2.3 miles southward up the road from location #3, you will come to Bunker Hill Road which leads west out of the Copper Creek district and returns you to River Road and Mammoth.

View of locations 1, 2, 3, and the Reliable Mine terraces from the road leading south from location 3.

*All G.P.S. coordinates taken at the collecting areas.

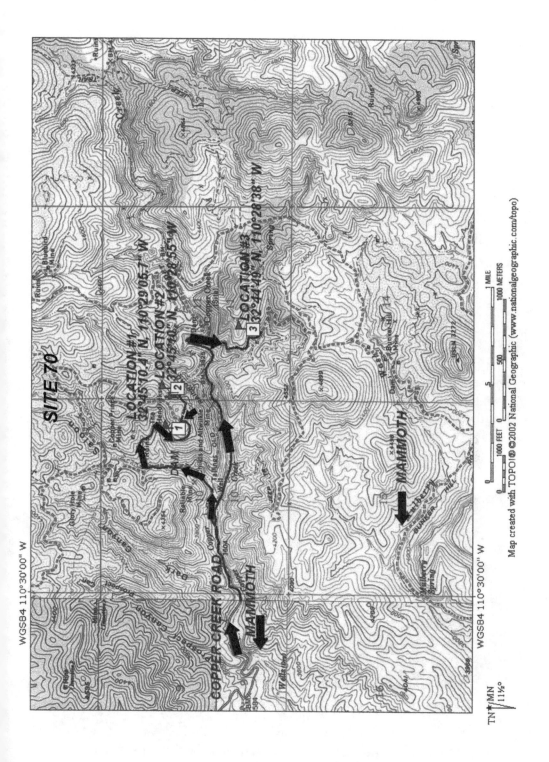

SITE 72

Selenite Roses
East of Saint David

Difficulty Scale: 4 – 4 – 5   Seasons: Fall, Winter, Spring

Global Positioning System Coordinates: 31° 51' 37" N, 110° 15' 18" W*

Geology: Holocene-Late Pleistocene Alluvial Clay, Sand, and Gravel

U.S. Geological Survey 7.5 Minute Topographical Map:
Land and McGrew Spring

FROM OCOTILLO ROAD AND 4TH STREET (U.S. Route 80) in Benson, drive southeastward toward Saint David 6.1 miles and turn right (south) on Apache Power Road. Go 3.6 miles to Desert Rose Road, turn right (west) and drive .5 mile to the T intersection of Desert Rose and Mountain View Roads. Turn left (south) and then make an immediate right turn (west) and continue driving toward the low bluffs ahead of you. After traveling .55 miles, you will come to the gas line road. Turn left here (south) and follow the gas line road .7 miles to

View of the low bluffs from the intersection of Desert Rose Road
and Mountain View Road.

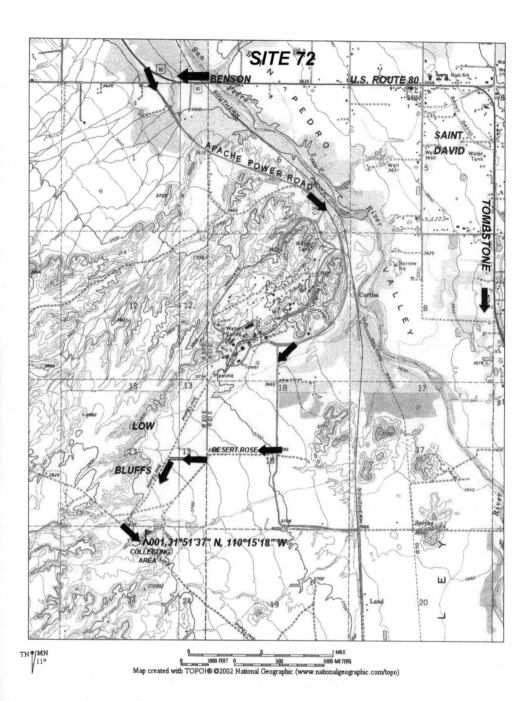

the beginning of the collecting area which extends southward for about .3 mile until you come to a locked gate.

This site is an Arizona classic. Well-known and actively collected for generations, it was thought to be nearly played out. Whereas in the past, collectors could hike over a wide area and discover selenite roses on the surface, in the sand, and among the ground clutter, collectors now have had to resort to digging into the clay banks that form the west bank of the San Pedro River Valley. The work is tough and the results are meager.

Less collecting has occurred less than a mile to the south down the gas line road. Therefore, the pickings there are easier and better although somewhat different from the rounder, sand encrusted, more earthy looking selenites typical of this area. Here, we have clearer, colorless selenite clusters with more transparency and vitreous luster. When you enter the collecting area, you are greeted by millions of small selenite shards shining in the sun like pieces of broken glass. The "roses" here are more like stars than roses having 3 or 4 long flat blades radiating from the center on a more or less horizontal plane. There are also some that are rounder and chunkier in form consisting of smaller crystal blades huddled together in a ball. Material is available on both sides of the road. However, the bluffs and valleys on the west side appear to be less collected.

*G.P.S. coordinates taken at beginning of collecting area.

SITE 74

Marble on Forest Road 689

Difficulty scale: 4 – 3 – 3   Seasons: Fall, Winter, Spring

Global Positioning System Coordinates: 31° 59' 57" N, 110° 00' 21" W*

Geology: Jurassic-Cambrian Metamorphic Marble and Quartzite

U.S. Geological Survey 7.5 Minute Topographical Map:
Dragoon and Knob Hill

FROM INTERSTATE 10 BETWEEN BENSON AND WILCOX, take exit 318, Dragoon Road, to the tiny town of Dragoon. From the railroad crossing in Dragoon, continue driving 1.55 miles east on Dragoon Road to W. Lizard Road which is Forest Road 689. Turn right (south) on W. Lizard, and follow it 2.3 miles up the canyon, past the small microwave antenna tower, to the first quarry. You can turn around and park here or you can drive up to the higher levels.

This quarry system is large enough to see from Interstate 10. It is a milky white scar in the green canyon. You will have a lot of area to cover if you desire

View of the marble quarry going up West Lizard Road (Forest Road 689).

to explore the entire complex. Most of the marble here is white although there is a small amount of black material here as well. There is probably enough marble here to reconstruct the Capital Building in Washington, D.C. Collecting is relatively easy. Marble chunks of all sizes from chips to large boulders are available. All you have to do is pick them up unless, of course, you are after a piece big enough to carve a life sized sculpture from. This marble contains a little willemite and will fluoresce a pale green and blue. It will also phosphoresce for a few seconds after you remove your UV device.

* G.P.S. coordinates are at first quarry.

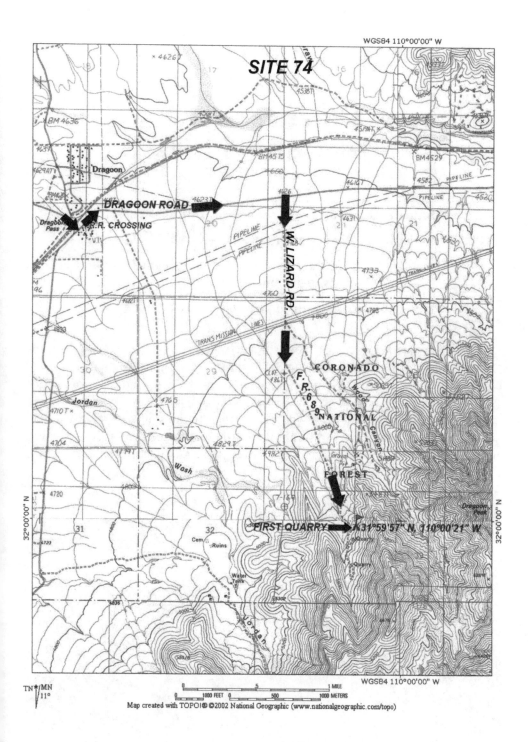

SITE 75

Minerals in the Tombstone Hills

Difficulty Scale: 5 – 6 – 4   Seasons: Fall, Winter, Spring

Global Positioning System Coordinates: 31° 42' 04.2" N, 110° 04' 26.8" W*

Geology: Early Tertiary-Late Cretaceous Igneous Granitic Diorite and Porphyry

U.S. Geological Survey Topographical Map: Tombstone

From the intersection of Old Charleston Road (S. Sumner Street) and W. Allen Street in Tombstone, go west on Old Charleston .7 mile to a gate on your left (south). As you proceed through the gate and up the dirt road, the Tombstone Hills will be directly ahead of you. To your left (east) back toward town you will see a tall telecommunications tower. Up the road to the south ahead of you will see a head frame. On your right, .5 mile up the road before you reach the head frame you will see a rough little road leading down into a gulch. This road will

Proceeding up the dirt road to the Tombstone Hills.

take you through the gulch and up around to the west slope of the big hill behind the head frame. If you follow the road southward beyond the head frame, you will come to a locked gate.

The collecting area extends from the tower to the east to the west slope of the big hill. It covers a distance of about 1 mile east to west. Below the tower, there is a large graded area, probably an old mine dump, with rubble piles in the gullies on both sides. In the vicinity of the head frame and in the excavations in the gully below it there are several mine dumps and tailings piles to explore. The largest and most challenging area is the west slope of the big hill. Wide and high and punctuated with numerous fenced off mine shafts, this location provides a huge amount of material for your prospecting pleasure.

The tombstone Hills are composed of early Tertiary to late Cretaceous granite and diorite rock containing copper deposits in near surface porphyry capped with limestone. Being highly mineralized, these hills were extensively mined for silver and other valuable metals. Consequently, the mine dumps provide fertile ground for mineral collectors. In addition to silver, antimony, copper, gold, lead, manganese, molybdenum, tellurium, zinc, and other substances were recovered here. The most prevalent and varied mineral of interest to rockhounds here is calcite. Travertine and aragonite are also present. The calcite is available in many different habits and forms. There are massive black, white, and clear deposits sometimes all together in the same rock. There are vugs and open seams sporting sharp, bright crystal formations. There are wide, shiny veins of bright clear calcite coursing through the dark, earthy matrix. Some rocks are honey combed with very delicate, frosty, spongy looking encrustations. These formations are stained with a wonderful variety of red, orange, and yellow tints and shades. Opaque snow white calcite orbs are scattered across the surface of some rocks like colonies of mold on old bread. When fractured, some rocks display radiating internal structures similar to wavelite. Others look like bundles of sticks and twigs. Dendritic stains decorate many calcite bearing rocks. There are also rocks a foot or more across containing wide, wavy, white and golden travertine onyx. Occasionally, you may be fortunate enough to discover a deposit of very pretty light blue onyx. The most striking specimens, however, are the puffy white and clear calcite formations piled on top of massive black calcite like pop corn balls. Most of this calcite fluoresces red, orange, or green. The dark colored material usually fluoresces red whereas the white calcite fluoresces orange and green. The fluorescence is not strong and flashy but, rather, tends to be more subtle and intermittent.

Throughout the entire collecting area from east to west, the ground is scattered with collectable material. You can systematically search the surface, dig through the many dumps, or bust up the bigger rocks to reveal the fresh specimens inside. Plan to spend a good deal of time here so you can cover the entire area completely and thoroughly.

Tombstone, "the town too tough to die," owes its past to silver mining. It owes its present existence to a gun fight. Prospector Edward Schieffelin founded the town by staking a silver claim in 1878. He named the claim Tombstone allegedly because a friend had warned him that if he prospected in Apache territory, all he would find would be his tombstone. He and his brother sold the Tombstone claim 2 years later for about 1 million dollars. The district became so successful that by 1882 the population is estimated to have grown to between 10 – 15 thousand people, making Tombstone the largest town in the Arizona Territory. In 1886, the mines closed after an underground fire destroyed the water pumps that had been installed to keep the constant underground water flow in check. Flooded, the mines remained inactive for the next 15 years until pumped out and reactivated by F.B. Gage who purchased most of the mines and formed the Tombstone Consolidated Mines Company. In 1909, the mines closed again because of a strike by the miners, the falling price of silver, and a railroad fuel car that was filled with salt water instead of fuel. The salt water ruined the pumping machinery and the mines flooded again. During its operating years, the Tombstone district produced over 32 million ounces of silver worth approximately $1.7 billion in today's dollars.

G.P.S. coordinates taken at the headframe.

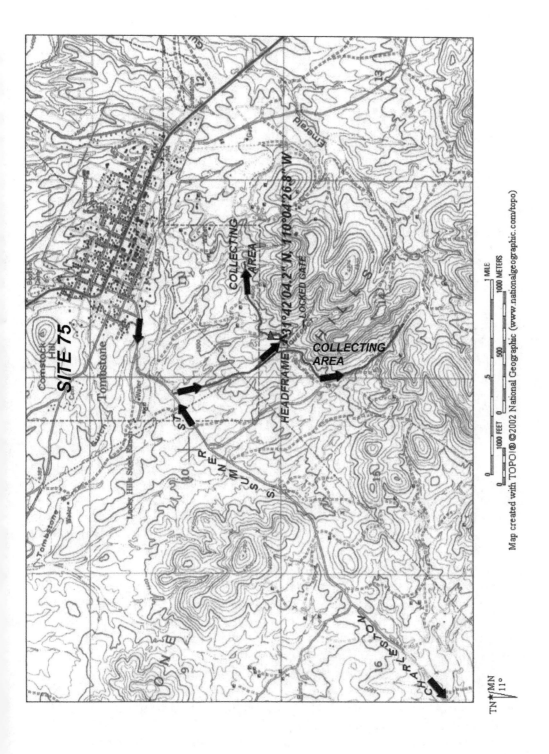

SITE 77

Quartz Crystal Druse at Gold Gulch

Difficulty Scale: 5 – 6 – 8   Seasons: Fall, Winter, Spring

Global Positioning System Coordinates: 31° 23' 52.2" N, 109° 51' 27.8" W*

Geology: Cretaceous-Late Jurassic Volcanic Rock

U.S. Geological Survey 7.5 Minute Topographical Map:
Bisbee and Bisbee NE

FROM BISBEE, TRAVEL SOUTH ON U.S. ROUTE 80 past the Lavender Pit to the traffic circle. Exit the circle on Bisbee Road. Follow Bisbee Road southward through the little towns of Bakerville and Warren to Arizona Road, also known as the Bisbee Junction Road. Bisbee Road becomes Douglas Avenue and then Ruppe Street before intersecting Arizona Rd. Turn right (south) on Arizona Rd. and drive .8 mile to S. Gold Gulch Road. Turn left (east) on S. Gold Gulch, travel 1.1 miles, and turn left (north) down the steep, rough road leading to the gulch below. Follow this road .3 mile to the point where 5 roads converge forming a star. There will be an old bullet riddled, upside down car on your left. Continue straight ahead across the star into the little canyon directly ahead of you. The collecting area stretches between the .2 and .4 mile points on the hillsides on the west side of the canyon. At the .2 mile point, you will see a newly blazed all terrain vehicle trail leading to crystal bearing rock outcrops a short distance up the hillside. You can see where previous rockhounds have been at work there. At the .4 mile point, you will come to the short road leading up to the Ivanhoe Mine. You can see the tailings pile before you get to the turnoff. About .1 mile up this road, you will come to a hairpin turn to the left. The hillside above you is covered with broken crystallized float. Collect the hillsides adjacent to the mine. There is nothing much of interest at the mine itself. Watch out for holes in the wash dug by gold prospectors as you drive up the canyon.

The host rock here is volcanic that intrudes the primary sedimentary Bisbee Group Formation. It is highly siliceous, making it hard to break. It is riddled with fairly large voids, vugs, and open seams which have become encrusted with bright, clear, small quartz crystal druse. Crystallized surfaces tend to be botryoidal forming clusters of BB-sized orbs.

View of Gold Gulch from S. Gold Gulch Road.

The crystal cavities that have not been exposed to the weather are clear, clean and exceptionally bright. Those that have been exposed are not only dirty, but are usually stained black with desert varnish or rusty red from the iron rich soil. Even when you break a rock open, you may find that iron contaminated water has seeped in and left a rust stained cavity. You can use oxalic acid to remove rust stains from exposed pieces. However, cleaned specimens are never as good as those that have remained unblemished. To harvest virgin specimens, you will need to employ a heavy sledge hammer, and expend a lot of sweat and energy to crack open the larger rock outcrops. Fortunately, the host rock, although very resilient, tends to split along the voids and seams resulting in drusey plates sometimes as much as a foot square. If too much matrix is attached to your specimen, you may be able to carefully trim it in the field with your rock pick or, if you have access to a large lapidary saw, you can cut it down to size when you get home.

*G.P.S. coordinates taken at newly blazed ATV trail.

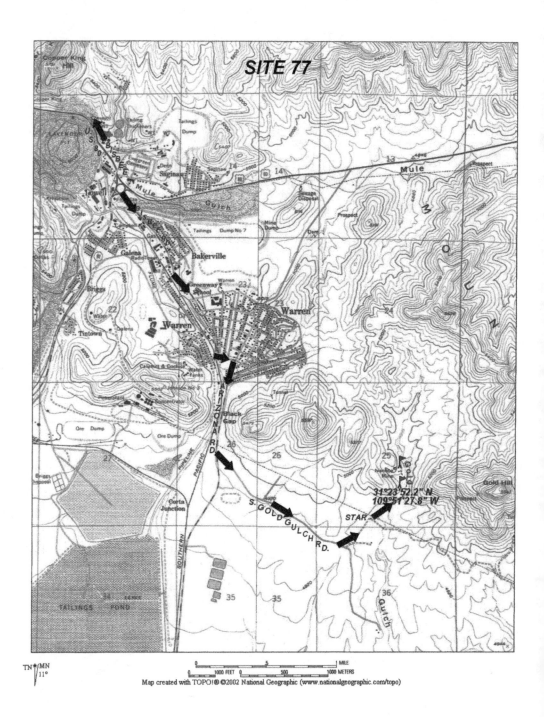

SITE 79

Porphyry at the W. A. Ranch Well

Difficulty Scale: 5 – 4 – 2   Seasons: Fall, Winter, Spring

Global Positioning System Coordinates: 32° 13' 45" N, 109° 00' 22" W*

Geology: Early Miocene-Oligocene Granite and Granodiorite

U. S. Geological Survey 7.5 Minute Topographical Map: Monk Draw

FROM THE INTERSECTION OF U. S. Routes 191 and 70 in Safford, go south on U.S. 191 26.2 miles and turn right (west) on a dirt road. If you are coming from Interstate 10, the distance to the turnoff is 9.8 miles. Drive 1.6 miles on the dirt road to a corral. Continue west around the corral passing through a green pipe gate. The W. A. Ranch Well is .3 mile farther on. You can park at the well and begin to collect in the wash and on the surrounding hillsides.

The parking area at the W. A. Ranch Well.

Pheonocrysts (Greek *phainein* meaning to show + *krystallos* for crystal) are partially formed feldspar crystals contained in igneous porphyry (Greek *porphyrites* literally meaning stone like Tyrian purple) rock. Colloquially called "Chinese writing stone" because its angular, geometric pattern resembles Chinese ideograms, the rock is a lapidary favorite. Fist sized pieces are scattered across the hillsides and in the washes immediately west of the well. You will begin to see them as soon as you park. Like most surface rocks that are exposed to the weather, the surface of these rocks is dulled and discolored. Inside, however, the feldspar pheonocrysts are clean and white and the igneous host rock is a pleasing pastel purple or gray-purple color. This material cuts well and takes a good polish.

G.P.S. Coordinates taken at the W. A. Ranch Well.

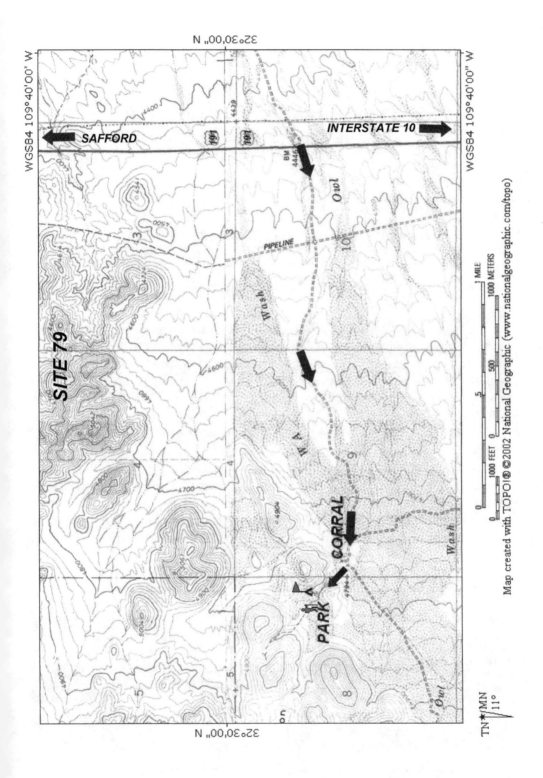

SITE 81

Chalcedony on Route to Coyote Spring

Difficulty Scale: 5 – 4 – 2   Seasons: Fall, Winter, Spring

Global Positioning System Coordinates: 32° 43' 43" N, 109° 17' 01.4" W*

Geology: Middle Miocene-Oligocene Volcanic Flows and Welded Tuffs

U.S. Geological Survey 7.5 Minute Topographical Map: Ash Peak

FROM U.S. ROUTE 70 BETWEEN Safford and Duncan, turn north on the graded dirt road opposite mile marker 366. The directions from here to the collecting area sound rather complex. But, if you pay close attention, you will find the trip to be easier than it appears. From U.S. Route 70, go .9 mile and bear right at the fork in the road. Go .7 mile to the second fork and bear left. After driving .9 mile, turn right. This road leads to Coyote Spring. The collecting area is between the .3 and .6 mile points on the hills and in the washes along this road.

Search the area on foot for unusual and aesthetically pleasing chalcedony formations. Be selective. Chalcedony is not a particularly rare or valuable gem material. It is really rather common stuff. Even so, it can be quite attractive. Chunky specimens that are pink or light tan in color, seem to glow from within, have a bright and shiny luster, and present an artistic shape are fun to discover and are desired by collectors. Leave the ordinary pieces in place. A good

The beginning of the collecting area on the road to Coyote Spring.

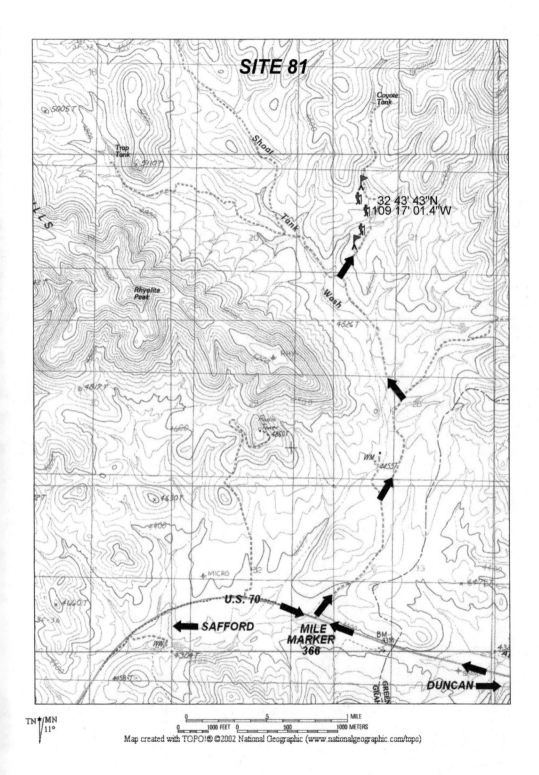

collecting rule of thumb is that if the specimen speaks to you, pick it up. If you have to think about it, best leave it alone. There are some pieces here that display dark burgundy colored sardonyx. Some of these pieces may contain fire agate. This collecting area is representative of many chalcedony occurrences in a large area bordered on the south by U.S. Route 70, on the east by the Gila River, and on the north and west by The Back Country Highway (the old Safford Road). Most of the real estate in this area is BLM land. Although, there are a few small patches of private property here and there. Agate, including fire agate, chalcedony roses, quartz crystals, a little petrified wood, and geodes are widely scattered throughout this area. The same is true of the Peloncillo Mountains to the south. If you search the area by driving the many primitive roads that traverse it and hiking its ridges and gullies you will certainly make new discoveries.

G.P.S. coordinates taken at the wash half way into the collecting area.

SITE 82

Geodes South of Ash Peak

Difficulty Scale: 5 – 4 – 5   Seasons: Fall, Winter, Spring

Global Positioning System Coordinates: 32° 43' 32" N, 109° 16' 40" W*

Geology: Middle Miocene-Oligocene Volcanic Rock

U.S. Geological Survey 7.5 Minute Topographical Map:
Ash Peak, Whitlock Mountains NE

THERE ARE TWO ROADS LEADING TO THIS SITE. Each one leads south from U.S. Route 70 between Safford and Duncan. If you are coming from Safford, Turn right on the maintained gravel road .1 mile beyond mile marker 363. Drive south .6 mile to the pipeline road. Turn left (east) and follow the pipeline road 4.1 miles and park. The collecting area is on your left (north). If you are coming from Duncan, travel 6.9 miles on U.S. Route 70 to mile marker 372. Turn left (south) on the un-maintained gravel road immediately opposite the mile marker post. Follow this road 2.6 miles to the pipeline road and turn right (west). Follow the pipeline road another 2.6 miles and park. The collecting area is on your right (north).

From the place where you park, it is a short, easy hike to the collecting area. Your destination is the rhyolite bluff at the north end of the low prickly pear studded ridge that runs northward from the road. Carefully pick you way through

The route northward from the pipeline road through the prickly pear to the geode site.

the prickly pear cacti until you reach the base of the rocky ledge. Along the way you will see a few geodes and geode pieces. Do not bother with these as there are literally thousands to choose from at the end of the ridge. As you near the bottom of the low ledge that forms the rhyolite bluff, you will find the ground covered with loose geodes that have weathered out of the rhyolite. The ground surrounding the bluff on all sides is littered with geodes. The rhyolite rocks and ledge faces are full of partially exposed geodes, ½ to several inches across, tightly packed and cemented together in the rhyolitic flow. Some are arranged in clusters like bunches of grapes.

The geodes here are typical of several other rhyolite flows in Arizona. At least half of them are solid rhyolite. Many of those that are hollow inside have small or off- center voids. Others are solid with chalcedony centers. The best specimens have thin walls and round hollow centers lined with chalcedony or quartz crystals. Most geodes here are smooth skinned and nearly perfectly round. The easiest way to collect is to search through the geode rubble fields surrounding the bluff for geodes that have naturally split in half. These halves will trim and polish very well. If you wish to collect whole geodes, then select those that are the lightest weight. These are more likely to be hollow. The heavier ones will surely turn out to be solid. Fortunately, you do not have to lug home large quantities of geodes and laboriously cut them open on your rock saw to find a few specimens worth keeping. The geodes at this location tend to split neatly in half if you are patient and tap them with your rock pick properly. Instead of smashing them into several pieces with one mighty blow, lodge the geode in a

The obsidian dike in the north side of the road cut on U.S. Route 70

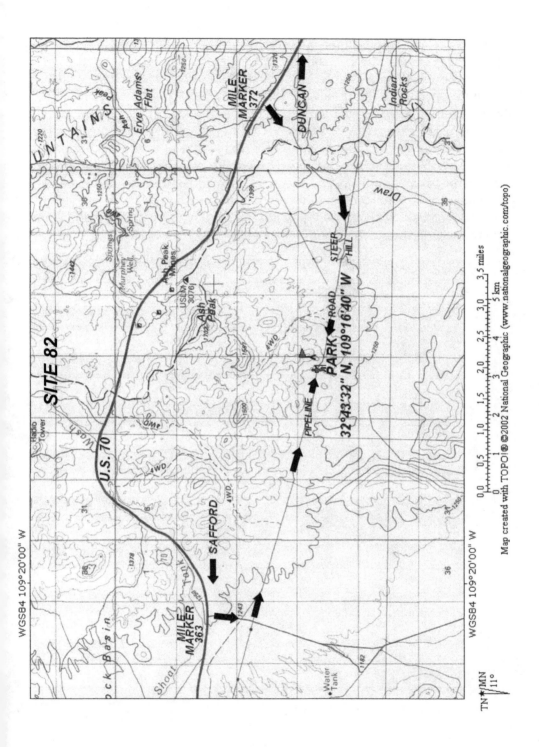

rock crevice and carefully tap it several times in the same spot. Usually this technique will result in the geode splitting evenly into two equal pieces.

As a bonus, there is an interesting side trip you may wish to take while you are in this area. Recent improvements to U.S. Route 70 between the turnoffs to the geode site have created a new road cut .8 mile west of mile post 365. At the east end of the road cut an impressive 10 – 20 foot wide vertical obsidian dike coursing through the volcanic country rock has been exposed. Do not try to park beside it in the narrow road cut. This would create a traffic hazard. Instead, park in the turnout on the north side of the road at the west end of the road cut. Rubble from the road cut containing obsidian was dumped here. You can also cross the road and walk up on top of the road cut on the south side of the road where you will find more exposed obsidian.

G.P.S. coordinates taken at parking area on pipeline road.

SITE 83

Opal on State Route 75

Difficulty Scale: 1 – 4 – 2   Seasons: All

Global Positioning System Coordinates: 32° 47' 54.7" N, 109° 09' 51.1" W*

Geology: Pliocene-Middle Miocene Sedimentary Sandstone, Limestone, Mudstone

U.S. Geological Survey 7.5 Minute Topographical Map: Sheldon

FROM THE INTERSECTION OF U.S. Route 70 and State Route 75 in Duncan, drive north on S.R. 75 7.7 miles and turn right (east) down into the parking area between the road and the fence. Park here, open the gate, and walk a few hundred yards up the wash to the limestone bluff to the north.

As you proceed up the wash toward the bluff, you will begin to encounter small opal chips and pieces that have washed down from the bluff ahead of you. The closer to the bluff you get, the more you will find. The opal here is the common type. It is bright, shiny, massive, and fractures in a conchoidal fashion like obsidian. Unfortunately, this is not your transparent rainbow colored gem quality or precious fire opal. Rather, it is opaque and generally porcelain white.

Looking north from the parking area toward the opal bearing bluff.

But, it will fluoresce a light pink color. Some specimens present a mottled brown or tan appearance. When you arrive at the bottom of the bluff, you will see the opal bearing stratum containing large 6 – 8 inch thick slabs of opal about 20 feet up the slope. The gullies running down the face of the bluff are clogged with opal chunks of various sizes that have broken off and fallen from above. Since this opal is very brittle, try to select pieces that are fracture free. If you wish to put it to lapidary use, work it slowly and carefully being sure to keep it cool. Excessive heat will cause it to crack.

G.P.S. coordinates taken at parking area.

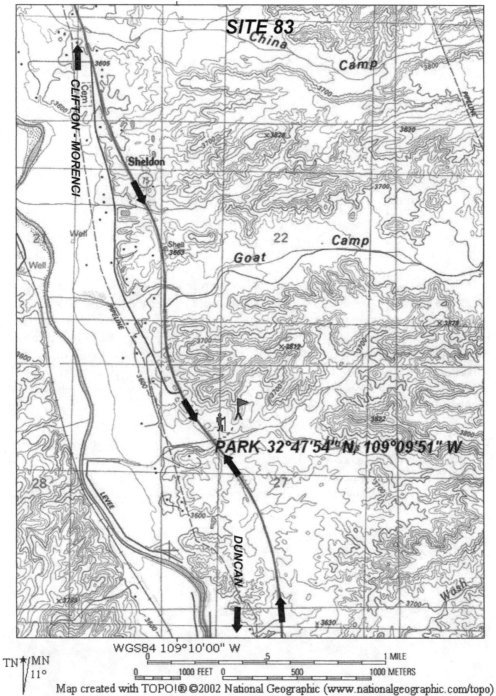

SITE 84

Minerals at the Carlisle Dumps

Difficulty Scale: 4 – 5 – 6   Seasons: Fall, Winter, Spring

Global Positioning System Coordinates: 32° 51' 04" N, 108° 57' 57.4" W*

Geology: Middle Miocene-Oligocene Lava Flows and Tuffs

U.S. Geological Survey 7.5 Minute Topographical Map:
Goat Camp Spring (NM)

ALTHOUGH THE SUBJECT OF THIS BOOK IS Arizona mineral collecting, there are four locations in New Mexico that are so close to the Arizona-New Mexico boarder as to warrant inclusion. These are Sites 84 through 87. To reach Sites 84 and 85, go to the intersection of U.S. Route 70 and State Route 75 in Duncan. From there, go 1.2 miles north on S.R. 75 and turn right (east) on Carlisle Road. After traveling 13.7 miles, the Carlisle dumps will come into view on your right as you approach a switchback in the road. Continue around the switchback on Carlisle Road another .2 miles and turn right (south). This road leads .2 mile into the Carlisle Mine pit and dumps on the north side of an east to west oriented ridge. Confine your collecting to the north side of the ridge. If you hike southward up and over the ridge, you will come to more recent excavations. This section of the Carlisle Mine is active and should be avoided.

The Carlisle Mine dumps cover a wide area and contain a variety of

View of the Carlisle Mine dumps from Carlisle Road.

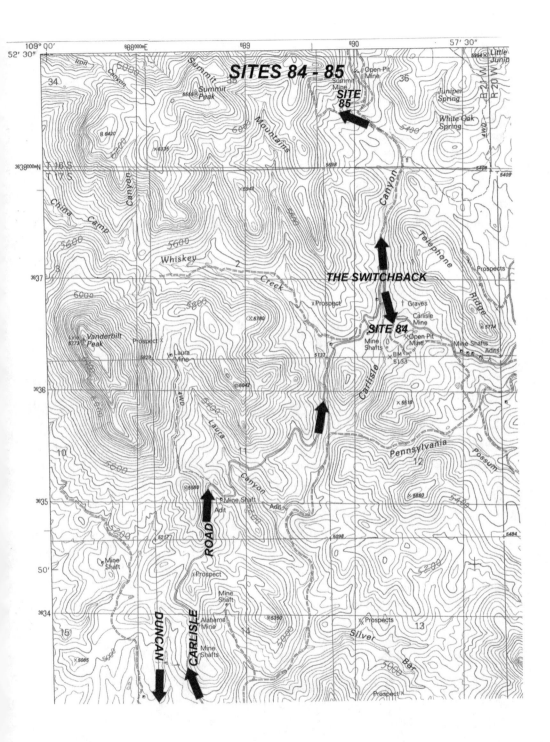

collectable minerals. At the east end is a shallow pit with tunnels penetrating the ridge. Do not attempt to enter these tunnels as they are dangerously unstable and flooded. The cliffs above the tunnel entrances are also severely fractured and undercut. This part of the pit may be of greater interest to the mineralogist then the collector because here you can observe the process of mineral creation occurring right before your very eyes. Around and above the tunnel entrances, conditions are such that the mineral chalcanthite is able to form. Chalcanthite is a water soluble copper sulfate secondary mineral that forms in oxidized zones of sulfur copper deposits. It has a remarkably bright, almost phosphorescent, blue color. It usually crystallizes into crusty or stalactitic forms. Here, it takes the appearance of piles of miniature popcorn balls randomly stacked on top of one another. Unfortunately, it is almost impossible to collect because it is so delicate. Even if you do succeed in harvesting a specimen intact, within a few days your museum piece will dry out, decompose, and crumble into white powder. The best way to capture this mineral is on film. Take a picture of it and leave it where it is for others to see. The adjoining cliff on the east side of the pit is better suited for collecting. Falling from the heavily fractured and unstable cliff above are rocks containing iron pyrite, quartz crystal, malachite, and chrysocolla. You may have to break open some of the larger rocks to expose fresher surfaces free from the yellow stain of oxidizing iron pyrite that is so prevalent here.

The remainder of the dumps west of the pit are hard rock mine rubble containing massive quartz, quartz crystal, a little light blue amethyst, and iron pyrite. The matrix is an attractive, pretty green, hard rock. Use a heavy hammer to break up the bigger rocks in search of crystalline vugs and pieces that reveal pleasing color patterns and mineral displays. These minerals encased in their green matrix make good cutting material.

In addition to mineral collecting, this location is a very interesting mining ruin. The remains of several stone buildings are evident throughout the dump area. On the slope above the dump is a large chimney and fire pit. Continuing up Carlisle Road beyond the dumps is a vault with a heavy reinforced iron door built into the side of a hill. At the top of the ridge on the left (east) is an intact head frame complete with engine, operator's shack, control levers, and cable descending into a vertical shaft. To the right and down the back side of the ridge the more modern equipment used to work the tunnels on the south side of the ridge is visible.

G.P.S. coordinates taken at the collecting area.

SITE 86

Dendrites near Steins, New Mexico

Difficulty Scale: 6 – 4 – 8   Seasons: Fall, Winter, Spring

Global Positioning System Coordinates: 32° 13' 45" N, 109° 00' 22" W*

Geology: Middle Oligocene Volcanic Thick Welded Tuff Promontories

U. S. Geological Survey 7.5 Minute Topographical Map:
Steins and Mondel (NM)

View of the road leading up to the bench below the high bluff.

SITES 86 AND 87 ARE JUST A FEW MILES across the New Mexico boarder along Interstate 10. You can actually see the first one from Arizona. To reach it, go east from San Simon on Interstate 10. Just before you cross the state line into New Mexico, you can see a high bluff in the Peloncillo Mountains on the north side of the Interstate. As you get closer to it and cross the state line, you can see the quarry workings and large concrete foundations on the hillside below the high bluff about .5 mile north of the highway. To reach it, take exit 3 off Interstate 10 to the "ghost" town of Steins. Drive through downtown Steins, about 200 yards, turn north across the railroad tracks, and drive .2 miles to the narrow, unmaintained, dirt road on your left (north-west). Follow this road .2 mile to a fork, bear left (west), and follow this curvy road 1.1 miles up to the quarry. You can drive up to the huge bench that is cut out below the cliff. Once on top, follow the road across the bench to the bottom of the highest point of the cliff.

Scattered all over the bench are piles of broken limestone ranging in size from dinner plates to pickup trucks. Although the Peloncillo Mountains are primarily volcanic, they also contain sedimentary strata. Dendritic (Greek *dendrophilious* meaning resembling a tree) stains are deposited in cracks in the rocks as water contaminated with dissolved manganese percolates through the limestone stratum. Sometimes mistaken for fossilized plant structures, the resulting arborescent (Latin *arbor* meaning tree) forms are really manganese that has recrystallized along the meandering water courses within the rock. Since rocks tend to fracture along existing cracks, there are many broken rocks scattered about displaying dendritic surfaces. Trimming these rocks down to a manageable size and shape is a hit and miss affair. This rock fractures conchoidally with no predictable pattern. Often, no matter how you strike them, they will split across the surface you are trying to harvest. Hopefully you will be able to fracture the rock in a way that produces a 1 – 2 inch thick plate displaying a face that looks like a pen and ink sketch of a forested landscape. If you feel really energetic, you can attempt to retrieve fresh, bright, sharp specimens by splitting open the larger boulders with sledge hammers and wedges.

*G.P.S. coordinates taken beside the concrete foundation below the bench.

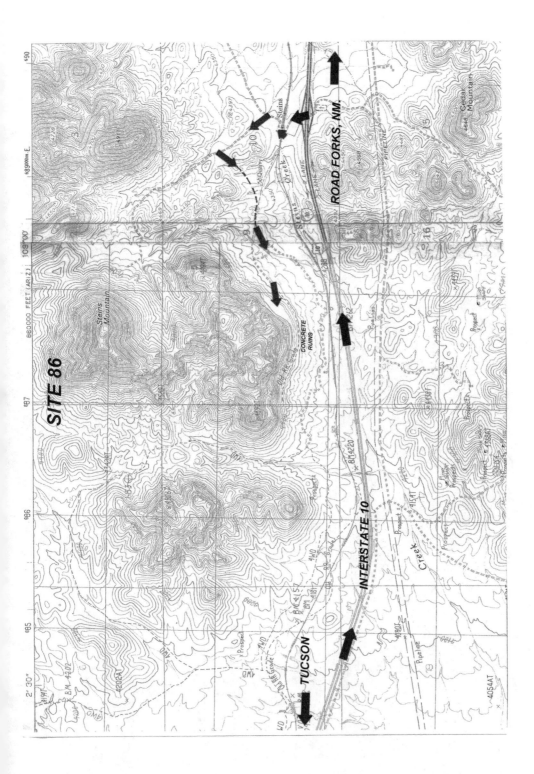

Color Plates

Key to Arizona Rocks Map

Each color on this map represents the predominant rock or sediment type within that colored area. The geology of Arizona is much more complex than the scale of this map is able to show. Smaller amounts of other type rocks are usually also present within each colored area.

All three rock types are present in Arizona. Igneous rock formed when molten rock (magma or lava) cooled and solidified. Lava is magma that reached the land surface. Sedimentary rock is made of compacted sediment. Metamorphic rock is igneous or sedimentary rock that was changed by high temperature and pressure, usually at great depths below the surface.

Igneous

Basalt: A hard black rock that formed when lava cooled. The lava was erupted less than 10 million years ago.

Mixed Volcanic Rocks: Cooled lava flows, ash, and related deposits more than 10 million years old. Some ash was red hot when it was erupted and the particles "welded" together.

Granite: Magma that cooled slowly and solidified beneath the land surface. This rock is now exposed at the surface because overlying rocks were removed by erosion.

Sedimentary

Silt, Sand, and Gravel: Eroded from mountains and deposited, mostly by streams, in valleys. Sand and gravel deposited by modern rivers are also included. Most of these deposits are less than 10 million years old and are weakly consolidated.

Sandstone: Deposited as sediment in a shallow sea, river delta, or by a large river system. Some mudstone is also present.

Limestone: Deposited as lime mud and the remains of marine animals in a shallow ocean. Some wind-blown siltstone and sandstone are also present.

Metamorphic

Schist and Gneiss: The processes of metamorphism may have taken place five miles or more beneath the land surface. Some rocks, under extreme temperature and pressure, were partially melted and stretched like taffy, producing banded rock called gneiss. Overlaying rocks were weathered and eroded exposing the once deeply- buried, metamorphic rocks at the surface.

Arizona Rocks Map

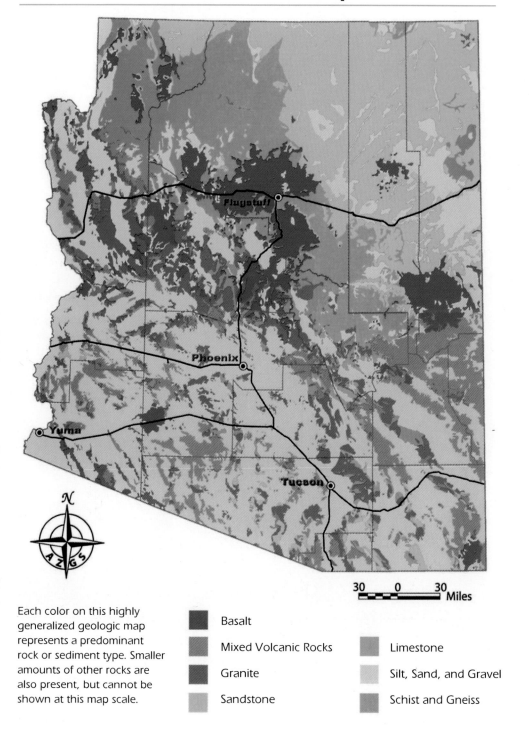

Each color on this highly generalized geologic map represents a predominant rock or sediment type. Smaller amounts of other rocks are also present, but cannot be shown at this map scale.

- Basalt
- Mixed Volcanic Rocks
- Granite
- Sandstone
- Limestone
- Silt, Sand, and Gravel
- Schist and Gneiss

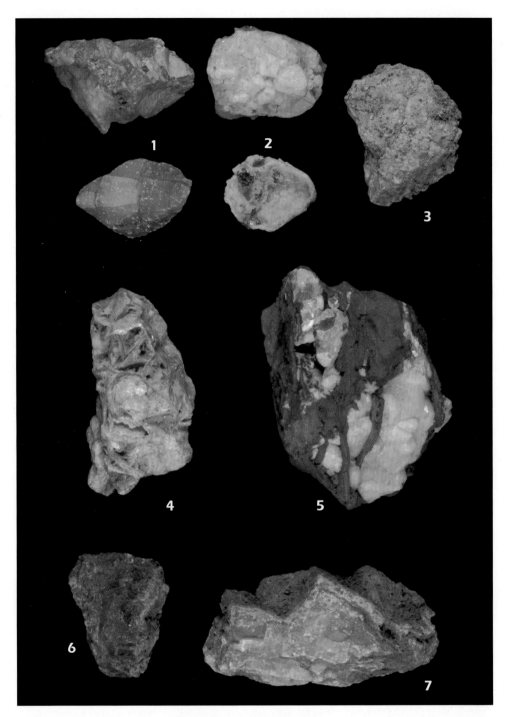

Plate No. 1: Fluorescents I

1. Site 31 – Calcite near Cottonwood Gulch. **2.** Site 38 – Geodes and chalcedony at the Maggie Mine. **3.** Site 40 – Calcite at the Scott and Blackpearl Mines. **4.** Site 41 – Calcite at the Black Silver Mine. **5.** Site 44 – Calcite at the Prince Mine. **6.** Site 43 – Calcite near Twin Buttes. **7.** Site 45 – Calcite north of Lime Hill.

Plate No. 2: Fluorescents II (Site 42 west of Morristown)

1. Location no. 1. **2.** Location no. 2 – The Newsboy Mine. **3.** Location no. 5 – The Big Spar Mine. **4.** Location no. 3. **5.** Location no. 4 – The Queen of Sheba Mine.

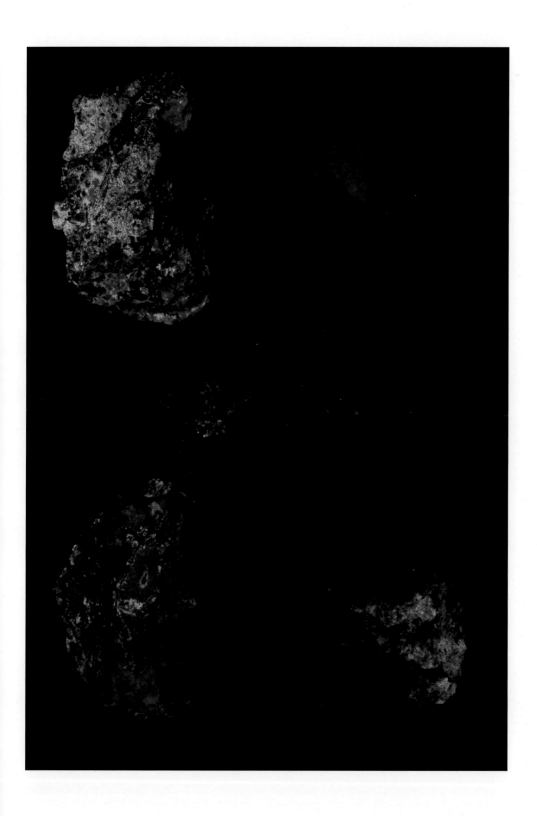

Plate No. 3: Fluorescents III

1. Site 63 – Fluorescents south of Cottonwood Canyon Rd. **2.** Site 65 – Fluorescents at the Ajax Mine. **3.** Site 66 – Fluorescent minerals south of Mineral Mountain location #1. **4.** Site 66 – Location #3. **5.** Site 66 – Location #5. **6.** Site 71 – Scheelite at the Tungsten King Mine.

Plate No. 4: Fluorescents IV

1. Site 78 – Calcite at the Crook Tunnel. **2.** Site 80 – Fluorescents along the San Francisco River Road. **3.** Site 85 – Calcite at the Summit Mine. **4.** Site 87 – Fluorescent minerals off U.S. Route 80. **5.** Site 88 – Scheelite at the Hilltop Mine. **6.** Site 39 – Fluorescent minerals at the Rawhide Mine.

Plate No. 5: Fluorescents V

1. Site 83 – Opal on S. R. 75. **2.** Site 29 – Calcite and chalcedony near F. R. 41. **3.** Site 27 – Glauberite pseudomorphs along Salt Mine Road. **4.** Site 11 – Fluorescents at Boyer Gap. **5.** Site 14 – Geodes east of the Ramsey Mine. **6.** Site 72 – Selenite rose east of St. David. **7.** Site 81 – Chalcedony on route to Coyote Spring. **8.** Site 74 – Marble on F. R. 689.

Plate No. 6: Fossils I

1. Site 2 — Mollusca in Nail Canyon. **2.** Site 3 — Crinoidea in Limestone Canyon. **3.** Site 22 — Marine fossils west of Jerome.

Plate No. 7: Fossils II (Site 21 — plant fossils at D. K. Well)

Plate No. 8: Fossils III

1. Site 23 — Crinoidea off State Route 89A. **2.** Site 28 — Rugosa near Chasm Creek. **3.** Site 48 — Mollusca at Battleground Ridge.

Plate No. 9: Fossils IV

1. Site 50 — Crinoidea off Forest Road 29A. **2.** Site 51 — Porifera and Brachiopodia along Forest Road 237. **3.** Site 54 — Rugosa and geode south of Control Road **4.** Site 55 — "Coral head" south of Forest Road 144.

Plate No. 10: Fossils V

1. Site 55 — Rugosa south of Forest Road 144. **2.** Site 56 — Crinoidea off Forest Road 29. **3.** Site 59 — Rugosa off Forest Road 138. **4.** Site 67 — Trace fossils at Dago Spring. **5.** Site 69 — Bivalva on State Route 77.

Plate No. 11: Fossils VI

1. Site 73 — Marine hash in French Joe Canyon. **2.** Site 76 — Gastropoda in the Mule Mountains. **3.** Site 89 — Gastropoda at the Cochise Mine. **4.** Site 90 — Crinoidea at the Willie Rose Mine. **5.** Site 68 — Marine fossils on route to the Grey Horse Mine.

Plate No. 12: Minerals I

1. Site 1 — Selenite at Dutchman Wash. **2.** Site 4 — Hematite at the BBC Mine. **3.** Site 5 — Cummingtonite east of the BBC Mine. **4.** Site 6 — Chrysocolla along the Rankin-Lincoln Ranch Road. **5.** Site 7 — Malachite at the Gold Hill Mine. **6.** Site 8 — Augite at the Green Streak Mine. **7.** Site 10 — Barite west of Plomosa Road.

Plate No. 13: Minerals II (Site 9 south of Plomosa Road)

1. Location no.1— Barite, hematite, and epidote **2.** Location no. 2 — Hematite and chrysocolla at the Mudersbach Mine. **3.** Location no. 1 — Granular hematite at the Phoenix and Yuma Mine. **4.** Location no. 1 — Selenite in the wash west of the Phoenix and Yuma Mine.

Plate No. 14: Minerals III

1. Site 12 — Alunite on Sugarloaf Peak. **2.** Site 13 — Bladed hematite at the Big Bertha Extension Mine. **3.** Site 15 — Chalcedony on Hull Road. **4.** Site 16 — Malachite at the Mammoth Mine. **5.** Site 17 — Copper minerals at Cunningham Pass. **6.** Site 18 — Copper minerals at the Bullard Mine. **7.** Site 19 — Amethyst at the Contact Mine **8.** Site 20 — Fluorite off Eagle Eye Road.

Plate No. 15: Minerals IV

1. Site 20 — Selenite off Eagle Eye Rd. **2.** Site 25 — Siderite at Copper Chief Mine. **3.** Site 24 — Jasper and hematite at Mingus Mtn. **4.** Site 27 — Glauberite pseudomorphs along Salt Mine Rd. **5.** Site 26 — Travertine at Empire Onyx Quarries. **6.** Site 49 — Chert at Battleground Ridge.

Plate No. 16: Minerals V

1. Site 29 — Calcite near Forest Road 41. **2.** Site 30 — Banded jasper off Table Mesa Road. **3.** Site 32 — Mica at Cottonwood Gulch. **4.** Site 57 — Marble off Ryan Ranch Road. **5.** Site 52 — Pegmatites along Black Pearl Mine Road. **6.** Site 58 — Opal and geodes at Corral Nuevo.

Plate No. 17: Minerals VI (along Signal and Alamo Roads)

1. Site 33 — Agate at Signal City. **2.** Site 34 — Banded agate and quartz crystals at the McCraken Mine. **3.** Site 35 — Chalcedony lined geodes near Keenan Camp. **4.** Site 36 — Fluorite at the Lead Pill Mine. **5.** Site 37 — Malachite at the New England Mine. **6.** Site 39 — Copper minerals at the Cactus Queen Mine.

Plate 18: Minerals VII

1. Site 46 — Obsidian on the Barry M. Goldwater Range — Area A. **2.** Site 47 — Geodes on the Barry M. Goldwater Range — Area B. **3.** Site 62 — Serpentine on Forest Road 189. **4.** Site 61 — Onyx on Forest Road 303. **5.** Site 60 — Chrysotile at the Phillips Mine. **6.** Site 53 — Malachite at the United States Mine. **7.** Site 60 — Verde antique on F. R. 189.

Plate No. 19: Minerals VIII

1. Site 64 — Amethyst at the Woodpecker Mine. **2.** Site 68 — Vanadinite at the Grey Horse Mine. **3.** Site 70 — Minerals from the Copper Creek District. **4.** Site 75 — Minerals in the Tombstone Hills.

Plate No. 20: Minerals IX

1. Site 77 — Quartz crystal druse at Gold Gulch. **2.** Site 79 — Porphyry at the W.A. Ranch Well. **3.** Site 84 — Minerals at the Carlisle Mine dumps. **4.** Site 82 — Geodes south of Ash Peak. **5.** Site 66 — Amethyst south of Mineral Mountain. **6.** Site 17 Hematite and copper minerals at Cunningham Pass. **7.** Site 86 — Dendrites near Steins.

Fossils

FOSSILS AND TRACE FOSSILS (Latin *fossilis* for "dug up") are the remains of ancient plants and animals and their activities that have been preserved in the earth's crust by a physiochemical process called fossilization. The science of fossils is called paleontology a Greek word meaning the study of life. At death, most organisms decompose and disappear—dust to dust. The miniscule number of flora and fauna who make it into the fossil record are celebrities. The deceased must be quickly and deeply buried upon passing to protect it from the bacteria that will consume it and the water that will dissolve it. Fossilization of soft tissue is almost nonexistent. Most fossilized organs are those made of hard tissue; woody tissue in plants and teeth, bones, and shells in animals. Therefore, collecting efforts may produce teeth, bones, mollusk and brachiopod shells, trilobite and crustacean exoskeletons, echinoderm parts, coral and bryozoan structures, and petrified wood. But, fossils of such delicate creatures as jellyfish and worms are almost nonexistent—bones to stones.

Fossil Formation

The study of fossilization is called taphonomy (Greek for grave). The first step in the fossilization process is the protection of the deceased from decay. This can occur in a number of ways. Freezing is the most desirable because it preserves the entire organism-hair, feathers, scales, skin, flesh, and internal organs as well as hard tissue. Well-known examples of extinct frozen creatures are ice age rhinoceroses and wooly mammoths that have become exposed in the arctic ice. Mummification, or desiccation, preservation by thoroughly drying the organism, also preserves tissue that would otherwise be lost to decay. The entrapment method was featured in the popular movie *Jurassic Park.* Since the release of this film, amber, petrified pine pitch, containing entrapped insects has become especially popular among collectors. In Poland, an entrapped wooly rhinoceros was recovered from a paraffin mine. The La Brea tar pits in Los Angeles contain the remains of animals and insects that were unfortunate enough to become stuck in the pit's gooey petroleum. Although somewhat dramatic, these fossilization processes are rare and usually yield fossils of organisms that lived only within the last few thousand years. You may possibly discover animal mummies in Arizona's hot dry climate, but not amber or frozen mastodons.

Once a dead organism has been properly protected from decay by one of the processes mentioned above, then true permanent fossilization can begin. If fossilization does not occur, then organic remains will decay when unfrozen, released from entrapment, or re-exposed to humidity. The most common fossilization methods are burial, mineralization, and formation of molds and casts. Burial is the simple act of quickly covering the organism with sediments that will prevent decay thus preserving it in more or less its natural state. The calcium-based shells of brachiopods and mollusks, for example, are sometimes found unaltered except for their loss of color. Peat bogs provide a favorable environment for preservation of both plants and animals. Wooden logs and stumps, animals, and even human bodies have been recovered from peat bogs, especially in Great Britain, relatively intact even after several thousand years. In Germany, 40 million year old lignite deposits have yielded wooden logs. Burial, however, usually results in the eventual mineralization of organisms. The most well-known example of this phenomenon is the great agatized tree trunks of Arizona's Petrified Forest National Park. Sorry, no collecting here. Mineralization occurs when water containing dissolved minerals soaks into plant or animal remains and replaces organic structures with stone. Frequently, the calcium in shells will be replaced with silica resulting in the creation of pseudomorphs. The primary replacement minerals in fossilization are calcium and silica provided by weathering volcanic ash. Other minerals that promote fossilization are carbon, phosphorous, and iron pyrite. In the carbonization process, all but the carbon in the organism decays, leaving paper-thin films of the organisms behind. These fossils look as if they had been burned or charred. Phosphatization prevents bones and teeth from dissolving as they usually do. Occasionally, an organism will be replaced by iron pyrite. These fossils have an unusually bright gemmy quality unlike most fossils that are dull and earthy in appearance. Often times, the fossil you find is neither the preserved nor replaced organism. It is, rather, a mold or a cast of the original. For example, if a brachiopod shell fills with sediment replacing the animal inside, the fossilized result is an internal mold of the inside of the shell. An external mold results when the shell leaves an imprint in a sediment that hardens into rock. If an external mold fills with sediment that hardens, then a cast is created.

The fossils discussed so far are all the remains, in one form or another, of actual organic structures. They are known as body fossils. There is, however another class of fossils called trace fossils. The study of trace fossils is called ichnology (Greek *ichnos* for footprint). As prehistoric creatures went about their daily business, they left evidence, traces, of their activities. The most exciting traces are tracks or footprints left by creatures such as dinosaurs, worms, and men that walked, hopped, jumped, crept, crawled, or slid through wet muddy ground that hardened into sedimentary rock thus preserving a record of their passing. The best example of this in Arizona is the dinosaur tracks near Tuba

City. Fossil dwellings in ancient sand, mud, and soil exist as burrows or borings. Trilobites laid eggs in pits called *cruziana* that are now part of the trace fossil record. In Nebraska the dens of extinct land beavers named *steneofiber* are found in buttes and bankings. Digestion creates trace fossil castings and coprolites. Castings are contorted strings of sand grains cemented together by ancient worms that ingested sand, digested the organisms that lived in it, and then excreted the sand. Coprolite (Greek *kopros* for dung and *lithos* for stone) is the collective term for the petrified droppings of fish, amphibians, reptiles, and mammals.

History of Paleontology

Fascination with fossils is nothing new. The human record shows that *Homo sapiens* have had an interest in the fossil record since ancient times. Along with gems and minerals, fossils have been discovered among the remains of prehistoric communal sites, dwellings, and graves. Fossil shark tooth pendants dating back to 3100 B.C. have been discovered in Egypt. In ancient Greece, the scholars Herodotus and Xenophanes collected, studied, and classified fossils. Writings and artworks from ancient civilizations reveal that fossils were mined, collected, marketed, studied, and displayed throughout the ancient world from Rome to China.

The discovery of fossils did not always foster scientific inquiry, however. Fossil bones often become the basis of legend and mystical interpretation that survives even to the present day. Man seems prone to create "explanations" to fill in the gaps in his scientific knowledge. One of the earliest recorded legends, about 675 B.C., involving mystical animals is that of the griffin. This eight-foot long creature is believed to be a product of both man's intellect as well as his imagination. It is pictured on ancient Greek pottery as having a cat-like body, wings, four heavily clawed feet, a fan-like protrusion at the back of its head, and a parrot-like beak. The story is that the Greek statesman and general Aristides first encountered griffins guarding the gold mines of the Scythian people at the foot of the Tien Shan Mountains along the border of China and Central Asia. Pure legend? Maybe. But, man's imagination has to be prompted by something. The Scythians lived around and traveled through the fossil-rich Gobi Desert where the fossilized bones of the dinosaur *Protoceratops* are found to this day. This creature had four heavily clawed feet, a long tail, an elevated frill at the back of its head, and a bird-like bill. What *Protoceratops* actually looked like in the flesh was, and still is, left to the imagination. The first known likeness of a griffin is a tattoo on the mummified skin of a Scythian who was buried in the Gobi Desert about 2500 B.C. Although the artist's rendering conforms fairly well to the fossil's skeletal form, the griffin's wings are, apparently, purely imaginary. That same imagination went to work again as our ancestors began to discover larger fossil bones. The discovery of ice age mammal bones had a particularly colorful effect on our literature and culture. They are the origin of giants like Cyclops the infamous one-eyed monster. Ancient literature is rich in geomystical creatures such as centaurs, tritons, satyrs, mermaids, sphinxes, minotaurs, dragons, and other amazing beings. Stories of these monsters are not only poetic, but prophetic as well. These fearsome creatures represented the evil unknown, the underworld. For much of our existence, we learned to fear and avoid the unknown. Seldom did we welcome and explore it.

During the Dark Ages, the monsters of the ancient world continued to flourish in medieval man's imagination. But, scientific study of fossils, like most everything else during that period, languished. Fossils were either reviled as works of the devil or simply dismissed as freaks of nature. As the renaissance dawned, modern paleontology grew out of the synergy generated by renewed scientific inquiry in botany, zoology, and geology. Leonardo da Vinci (1452-1519) recognized that fossils were indeed the vestiges of ancient flora and fauna. The writings of Agostino Scilla (1629-1700) and Fabio Colonna (1567-1650) compared fossil structures to those of living organisms. Father Nicolas Steno (1631-1686) formulated the geologic laws of successional sedimentary strata and original horizontality. The naturalist Carl Von Linné (1707-1778) devised the systematic binomial nomenclature by which we classify and name both living and extinct species. During the nineteenth and twentieth centuries, natural scientists (botanists, zoologists, geologists, paleontologists, and others) performed extensive field explorations. Discoveries were made and put on public display in museums around the world. This sparked a renewed interest in mineral and fossil scholarship and recreational collecting that persists to the present day.

Stratigraphy (Latin *stratum* for covering)

Knowing what fossils are and how they came to be is interesting enough. But actually finding them is another matter altogether. The key to successful fossil foraging is knowing what type of rock from which geologic time period to look for. Since the beginning, some 4.5 billion years ago, the earth has been in a state of slow but constant geologic change. Episodes of geographic, geologic, and climatologic change occur one after the other like the acts and scenes of a play. Collectively, these events are known as the Geological Time Scale and are divided into eons, eras, periods, and epochs. The history of this geologic saga is literally written in stone.

The rock cycle arranges the earth's crust into types of rock layers: igneous, sedimentary, and metamorphic (see page 35). The sequence of strata is called the geologic column. The study of the geologic column is called stratigraphy. Miners create examples of the geologic column by tapping into the earth's crust and extracting core drills. You may discover these in the form of rock cylinders discarded on abandoned mine dumps. The most dramatic revelation of the column, however, is the Grand Canyon where you can view the geologic history of Arizona like the pages of a book and the volumes of an encyclopedia.

Understanding the basic principles of stratigraphy will help you interpret the geologic environment you face on your fossil hunts. These principals provide the basis for dating the appearance, evolution, and eventual extinction of fossilized organisms. By applying these principals, strata can be identified, classified, and mapped over wide geographic areas.

1. The Principals of Successional Sedimentary Strata and Original Horizontality. Simply stated, sediments are deposited one atop another in a flat horizontal attitude with the lowest or first layer being the oldest and the highest or last layer being the youngest. This principal is the basis for determining the age or succession of strata relative to each other.

2. The Principle of Cross-Cutting. A geologic formation that intrudes or cuts across another layer of rock is younger than the rock intruded upon. This principle is the basis for dating faults, folds, and other disturbances.

3. The Principal of Inclusion. Rocks that contain fragments of other rocks must be younger than the rocks they contain. This principal is the basis for determining the age of conglomerates and the source of sediments.

4. The Principal of Fossil Succession. Organisms evolve and follow one another in time and in a non-repetitive sequence. This principal is the foundation of biostratigraphy—the classification and dating of rock strata based on fossil content.

5. The Principal of Uniformitarianism. The present is key to the past. Geologically, present day environments (deserts, swamps, ocean bottoms, mountain ranges, etc.) display the same characteristics and create the same conditions now as they did in the past. Biologically, organisms lived and organic structures functioned the same then as they do now. This principal is the basis for interpreting fossils and their environments.

Based on the above laws, natural science developed a chronological calendar of Earth's geological development from the beginning to the present. The biological calendar is merged with the geological record using the principal of fossil succession. Knowing the time periods in which certain fossilized organisms lived, the stratum they are found in can then be dated. The fossil calendar contains several cycles of sequential development wherein life forms emerge, evolve, and then become extinct because of environmental changes. New life forms evolve from the few organisms that are able to successfully adapt to their new environment, a new epoch begins, and the cycle repeats itself. See Table 14, page 258, for select Arizona sedimentary formations.

The calendar of life is divided into eons spanning Earth's 4.5 billion year

TABLE 14
Selected Arizona Sedimentary Formations

Formation	Composition	Period	Arizona Location
Amole	Shale, sandstone	Cretaceous	Tucson Mountains
Bidahochi	Conglomerate	Cenozoic	Twin Buttes
Bisbee Group	Conglomerate, shale, sandstone	Cretaceous	Bisbee
Bright Angel	Shale	Mesozoic	Grand Canyon
Chinle	Conglomerate, shale	Middle-Late Triassic	Chinle Valley
Concha	Limestone	Permian	Chiricahua Mountains
Claflin Ranch sandstone	Mud, conglomerate,	Late Mesozoic	Silverbell Mountains
Coconino	Sandstone	Permian	West Central Arizona
Cow Springs	Silty sandstone	Upper Jurassic	Black Mesa
Dakota	Sandstone	Cretaceous	Tuba City
Escabrosa	Dolomite, chert, limestone	Early Mississippian	Superior, Huachuca Mountains
Hermit	Shale	Permian	Limestone Canyon
Hoquilla	Limestone	Permian	North of Portal
Kaibab	Limestone	Middle Permian	Kaibab Forest
Kanab	Limestone	Mississippian-Pennsylvanian	Kanab Canyon
Mancos	Shale	Cretaceous	Grand Canyon
Martin	Limestone, dolomite, sandstone	Middle Devonian	West Central, Jerome

Formation	Composition	Period	Arizona Location
Mescal	Limestone	Proterozoic	Superior
Naco Group	Limestone, shale	Permian-Pennsylvanian	Naco
Redwall	Dolomite, chert, limestone	Early Mississippian	Coconino National Forest, Jerome
Silver Bell	Mud, sandstone, conglomerate	Late Mesozoic	Silver Bell Mountains
Sonoita Group	Shale, sandstone	Late Cretaceous	Sonoita Flats
Supai	Limestone, shale, sandstone	Permian	Clarkdale, Limestone Canyon
Tapaets	Shale, sandstone	Late Cretaceous	Black Mesa
Toroweap	Sandstone, limestone	Pennsylvanian	Mogollon Rim
Verde	Limestone	Pliocene-Miocene	Verde Valley

history. The period before life appeared on Earth is named the Archean eon, the period of "deep time", between 4.5 and 2.5 billion years ago. This was followed by the Proterozoic or Precambrian eon between 2.5 billion and 570 million years ago which saw little life save a few one-celled organisms such as bacteria, algae, and rudimentary jellyfish. The third is the Phanerozoic eon between 570 years ago and the present. The Phanerozoic eon is divided into eras. The first is the Paleozoic (Greek for ancient life) era beginning about 570 million years ago. It is known as the Age of Marine Invertebrates, Fishes, and Amphibians. The second is the Mesozoic era (Greek for middle life) beginning about 245 million years ago. It is known as the Age of Reptiles. And finally, is the Cenozoic era (Greek for recent life) beginning about 66 million years ago. It is the Age of Mammals that we are presently in. Eras are then subdivided into periods that are further subdivided into epochs. See Table 15, page 260.

Because organisms have a better chance of fossilizing at the bottom of bodies of water than on dry land, the best environment to conduct fossil searches is in marine sedimentary rock strata. Pressure and chemical action cause underwater sediments to harden into rock. Gravel becomes conglomerate, sand becomes sandstone, mud becomes shale, and organic calcium carbonate becomes limestone. Some of the Proterozoic eon and Paleozoic era species such as algae, plants, invertebrates, fishes, and amphibians that lived in ancient seas became fossilized in these sedimentary strata.

TABLE 15
Chronology of Life on Earth

Eon	Era	Epoch	Period – Millions of Years Ago	Organism Appearance and Extinction
PHANEROZOIC	CENOZOIC	Holocene, Pleistocene.	Present. *Quaternary*	*Homo sapiens*, ice age mammals, whales, horses.
	The Age of Mammals.	Pliocene, Miocene, Oligocene, Eocene, Paleocene	1.6 M *Tertiary*	Hominids, primitive mammals, large sharks, grasses, oreodonts.
			66.4 M	Extinction of dinosaurs and 75% of species.
	MESOZOIC	Late and Early	135.0 M *Cretaceous* (Latin – chalk)	First primates, snakes, flowering plants, birds.
	The Age of Reptiles	Late, Middle, Early	144.0 M *Jurassic* (Swiss mountain range)	Dinosaurs, first mammals, frogs, conifers
			208 M	75% species extinction.
		Late and Early	225 *Triassic* (Latin – 3 layers)	First dinosaurs, cycads, gingkoes, crocodiles.
			245 M	Largest extinction. 95% of species lost.
	PALEOZOIC	Late and Early	280 M *Permian* (A Russian province)	Large reptiles, amphibians, pine trees.
	The Age of Amphibians	Late, Middle, Early	310 M *Pennsylvanian* (carboniferous)	First reptiles, insects, spiders, coal swamps.
		Late and Early	345 M *Mississippian* (carboniferous)	Crinoids, amphibians
			367 M	80% species extinction.

Eon	Era	Epoch	Period – Millions of Years Ago	Organism Appearance and Extinction
P H A N E R O Z O I C	*The Age of Fishes*	Late, Middle, Early	400 M Devonian (Devonshire, England)	First amphibians, sharks, ferns, rushes, club mosses.
		Late, Middle, Early	430 M Silurian (A pre-Celtic tribe)	First land plants and animals.
			438 M	85% species extinction.
	The Age of Marine Invertebrates	Late, Middle, Early	500 M Ordovician (A Celtic tribe)	First corals, sea stars, fresh water fish.
		Late, Middle, Early	570 M Cambrian (Latin – Wales)	First, mollusks, trilobites, brachiopods, and fish.
	PRECAMBRIAN PERIOD (Proterozoic Eon)		570 M– 4,500 billion years	First life appears. Arizona's oldest rocks. Blue and green algae, bacteria, one-cell organisms.
	ARCHEAN EON		Deep Time	No Life.

Sedimentary strata differ according to the environments in which they were deposited. Consequently, different types of flora and fauna will characterize different strata. Strata that are laterally contiguous, have similar lithologic characteristics, and show clear upper and lower boundaries, are designated as formations. Formations are the basic unit of geological mapping. They are usually named after a prominent geographical feature or area such as the Redwall Limestone named after the red cliffs of the Grand Canyon and the Naco Formation named after the little border town south of Bisbee. Similar formations may be combined into groups. Formations are subdivided into members and beds. Once identified, formations can then be drawn on a geologic map. Each rock formation is the creation of the environment that prevailed in a given location at a particular time. Geographic mapping, then, reveals how series of environments are related on a panoramic scale over large geographic areas. It is important to remember, however, that the same type of formation, Redwall Limestone for example, may vary in age from location to location. As seas invaded and subsided during different geologic times, the same environments, and therefore the same formations, were recreated numerous times. Regardless of the era, beaches deposit sand, offshore water deposits silt and mud, and deep seas precipitate limestone. As seas advance and retreat over the landscape, the location of these environments changes thus leaving the same formations at different locations at different times.

Fossil Classification

We know what fossils are, how they are created, and where to search for them. But once we catch them, what shall we name them? Enter Swedish naturalist Carl von Linné. His book *Systema Naturae* established the system of biological nomenclature for both living and fossilized organisms that is used worldwide today. According to Von Linné's system, all living things belong to one of two kingdoms: Plantae (plants) or Animalia (animals). The classification system then subdivides plants and animals into categories. These categories, arranged in descending order from the most general to the most specific, are: phylum, class, order, family, genus, and species. These classification categories are synthetic groupings based upon recognized similarities or general semblances. For example, mankind is of the Kingdom Animalia, Phylum Chordata (vertebrates), Class Mammalia, Order Primate, Family Hominidae (resembling man), genus Homo (man), and finally Species *Homo sapiens* (wise man). Members of the same species are able to produce fertile offspring. The genus category recognizes similar biological appearances. Lions, tigers, and house cats are of the same genus. Linné's nomenclature is an international system of standard terms based on the Latin language. His system assigns single word names to each category except species. The species category is binomial having two names. The first is the genus name and is capitalized. The second name may be the person who discovered the organism, the geographic location where it was originally found, or a distinguishing characteristic of the organism. The second name is Latinized and not capitalized. Man is genus *Homo* and species *sapiens*. His house cat is *Felis domesticus*. When written, both the genus and species names are italicized.

It is not necessary to become a linguist or a fossil classification expert to enjoy the hobby of fossil collecting. Some collectors become amateur paleontologists, finding a more scholarly approach to be more interesting and rewarding. Others are content with a more generalized approach using just the phylum or common informal names to identify their finds. Following is a list of the most commonly found marine invertebrate fossil phyla found in Arizona. Common and informal names are included since they are often heard in fossil collecting circles. Most fossils are of extinct species. However, in some cases, younger species of the same phylum exist today.

1. Phylum Porifera. Latin *porfer* for "pore bearers". Common name: sponges. Porifera have existed since the Precambrian era. They are multicellular, not symmetrical, and live singly or in colonies. As juveniles, they drift in the water, but attach to the bottom in adulthood. Sponges come in three types. Calcareous are the simplest having calcite or aragonite skeletons. The glass type has silicon spicules—long slender supports that serve as skeletons. And, the third type is the siliceous sponges, which are the most numerous and complex. Sponges have no nervous system or internal organs. They breathe and feed by drawing water in and out through holes on the surface of their bodies.

2. Phylum Coelenterata. Latin for "hollow gut". Common name: corals. This Phylum also contains jellyfish and sea anemones. Corals are in Class Anthozoa. Of the three sub-classes of Anthozoa, only one, Zoantharia is known to the fossil record. Orders under Zoantharia are Rugosa, Tabulata, and Scleractina. The Rugosas are known as horn corals, have a calcareous skeleton, and lived from the Ordovician to the Permian periods. The surface of their exoskeletons are marked by wrinkles. The Tabulate corals lived in the same periods as the Rugosas. They are colonial, forming chains, nets, tubes, and honeycomb groups. Scleractineans first appeared in the Triassic period. These are the corrals that laid the foundation for our present day coral reefs. Corals reproduce sexually and asexually by budding.

3. Phylum Bryozoa. Greek *bryon* for moss. Common name: moss animals. Bryozoans are aquatic, living in both salt and fresh water, and are similar in appearance to corals but, are much more complex. They first appeared in the Ordovician period and continue to flourish today. They are called moss animals because they have a fuzzy appearance similar to moss growing on the ground. They have internal digestive organs, a calcareous skeleton, and live in colonies. The best way to tell a Bryozoan from a coral fossil is to compare the pores. Bryozoan pores are pin-sized. Coral pores are larger and easily seen with the naked eye.

4. Phylum Brachiopodia. Latin *brachium* + Greek *podpous* for "arm foot". Common names: Brachiopod, lamp shell, brachs. Brachiopods live on the sea floor either attached to or simply resting on it. There are two types, articulate and inarticulate. The inarticulate are the oldest dating from the Cambrian period. They have two unhinged shells composed of chitinous-phosphate material. The articulates came later in the Cambrian Period. They have two hinged calcite shells. Brachiopods are often mistaken for mollusks because both animals have two opposing shells. But, the two are not even in the same phylum. When mollusk shells are separated and placed side by

side, they are mirror images of each other. Brachiopod shells are not. Brachiopod shells are different from one another. The bottom shell is more bowl-shaped and the top shell is shaped somewhat like a lid.

5. Phylum Mollusca, Class Bivalva. Latin *molluscus* for soft. Common names: clams, scallops, mussels, oysters, shellfish. Bivalves first appeared in the fossil record in the early Cambrian period and are the dominant hard-shelled marine invertebrate today. Mollusks recovered and evolved more successfully than brachiopods after the great extinction at the end of the Permian period. They live on the sea bottom, burrow into the sediment below it, attach themselves to natural and man-made objects, and some can swim in the water column. Most species have a foot for burrowing or boring. Except for oysters and rudists, bivalves have two bilaterally symmetrical hinged shells. Large colonies of oysters and rudists form reefs.

6. Phylum Mollusca, Class Gastropoda. Latin for "stomach foot". Common name: snails. Gastropods are the most diverse class of mollusk fossils with over 15 thousand species. They have two shell configurations, coiled and spiraled. Both types are made of aragonite. The most common shell design is the helical spiral, where the shell is coiled about a vertical axis. They first appeared in the early Cambrian period and have adapted so well that today there are over 85 thousand species inhabiting salt water, fresh water, and terrestrial environments worldwide. Unlike bivalves, gastropods have a true head containing eyes and other sensory organs. Movement is accomplished by creeping or sliding on a flattened foot.

7. Phylum Molluska, Class Cephalopoda. Latin *cephalad* + Greek *podpous* for "head foot". Common name: nautilus, ammonite, horn shell. Nautiloidea and Ammonoidea are subclasses of Cephalopoda. Nautiloidea dates from the lower Ordovician period and still exists today. Ammonoidea appeared during the lower Devonian period and became extinct in the upper Cretaceous. These creatures have sophisticated heads and impressive brains. They have feet-like tentacles positioned in front of their mouths for moving about and catching prey. Their eyes are highly developed. They can jet through the water by squirting pressurized water out of their bodies. Most fossils are internal molds since their aragonite shells easily dissolve away. Arizona nautiloids are contained in Paleozoic marine formations. The name ammonite is Greek for Ammon's stone. The ornamented wrinkled whorls of the animal's shell resemble the ram's horns of the Egyptian god Ammon. Not all ammonites are coiled, however. A few are linear.

8. Phylum Echinodermata. Greek *echin* + *derma* for "sea urchin skin". Common name: sea lilies, crinoid stems. Echinoderms have exoskeletons of spiny porous calcite plates, which normally have pentameral or five-rayed symmetry. Internally, they have a complex water-vascular plumbing system.

TABLE 16
Fossil Environments

Organism	Phylum / Class	Period	Environment
Corals	Coelenterata / Anthozoa	Ordovician to present	Marine: warm, clear, shallow water. Mud, Sand, limestone bottom.
Sponges	Porifera	Cambrian to present	Marine: clear, shallow water.
Bivalves	Mollusca / Bivalva	Cambrian to present	Marine and fresh water. Burrow in sand and mud and attach to bottom.
Gastropods	Mollusca / Gastropoda	Cambrian to present	Marine, fresh water, and terrestrial.
Nautiloids and Ammonites	Mollusca / Cephalopoda	Ordovician to present. Ammonites are extinct.	Marine: swim, float, dwell on bottom.
Trilobites	Arthropoda / Trilobita	Cambrian to Permian. Extinct.	Marine: deep and Shallow water. Swim, burrow, crawl.
Moss Animals	Bryozoa	Ordovician to present	Marine and fresh water shallows.
Brachiopods	Brachiopodia	Cambrian to present	Marine: tidal zone to 18,000 feet. Bottom dwellers.
Crinoids and blastoids	Echinodermata	Ordovician to present	Marine. Mobile and attached to bottom.

This system extends outside the body as tube feet that enable the animal to breath, feed, and move about. They appeared in the early Paleozoic period and are abundant in seas worldwide today. Two members of this phylum are abundant in the Arizona fossil record: crinoids (Greek *krinoeides* for lily) and blastoids (Greek *blastos* for bud). These animals are easily mistaken for plants because they can attach to the seabed with feet resembling roots, have a long segmented column resembling a stem or stalk, and they have a calyx or head that looks like a flower bud. Crinoids have arms protruding from their calyx. Blastoids do not. Disarticulated pieces, commonly referred to as crinoid stems, rather than entire animals are usually found. Blastoid calyxes looking like petrified flower buds or nuts are sometimes found.

9. Phylum Arthropoda. Greek *arthra* + *podpous.*, Class Trilobita. Greek *trilabos* for "three-lobed". Common Name: trilobites. The Phylum Arthropoda contains trilobites, the fossil we love to find, crustaceans, the shrimp and lobsters we love to eat, insects, the bugs we love to hate, as well as millipedes, horseshoe crabs, barnacles, and others. Members of this phylum inhabit every environment from aquatic to terrestrial to aerial. Trilobites came in over 15 thousand genera and several thousand species that lived in ancient seas from the lower Cambrian to the middle Permian period. Trilobites had a flexible protective exoskeleton that bent latitudinally at the cephalic (head), the thorax (body), and at the pugilism (tail) regions. The thorax was also divided into three lobes longitudinally. The extra flexibility provided by these lobes enabled the creature to travel nimbly over the ocean bottom and roll up in a ball like an armadillo when threatened. Like modern day crustaceans, trilobites molted. Consequently, a single animal could be responsible for several fossil remains. Trace trilobite fossils also exist in the form of burrows, nests, and tracks.

The Fossil Collecting Sites

SITE 2

Mollusca in Nail Canyon

Difficulty Scale: 2 – 4 – 4   Seasons: Spring, Summer, Fall
Global Positioning System Coordinates: 36° 40' 20.1" N, West 112° 22' 05" W*
Geology: Permian Sedimentary Kaibab Limestone Formation
U.S. Geological Survey 7.5 Minute Topographical Map: Toothpick Ridge

The entrance to the Nail Canyon fossil site at the intersections of FDRs 423 and 642.

FROM FLAGSTAFF, DRIVE NORTH on U.S. Route 89 as if you were going to the north rim of the Grand Canyon, which you should definitely do since you will be so close to it. U.S. 89 will take you to U.S. Alternate Route 89, which crosses the Colorado River at Navajo Bridge south of Parker and Glen Canyon Dam. Proceed west on Alternate 89 to the intersection of State Route 67, The North Rim Parkway, at Jacob Lake. At Jacob Lake, stop at the Kaibab National Forest Ranger Station and ask for the Snake Gulch–Kanab Creek Trail #59 handout that shows the way to Nail Canyon and the surrounding area. From the ranger station, go south on Route 67 .2 mile and turn right (west) on Forest Development Road (FDR) 461. Go 5.5 miles to West Side Road (FDR 22) and turn left (south). Drive

2 miles and turn right (west) on FDR 423. Proceed 1.3 miles and turn right on FDR 642, the entrance to Nail Canyon. FDR 642 goes into Nail Canyon for about three miles before ending at the trailhead 59 parking lot.

The fossils here are primarily bivalves and gastropods. At the end of FDR 642, several large fossiliferous Kaibab Limestone Formation rocks define the boundaries of the parking lot. **Do not try to collect these**. Instead, use them as index rocks to tell you what to look for elsewhere in the canyon. Search along the base of the cliffs on both sides of the canyon. Lots of debris has piled up there from the eroding dark limestone sedimentary layers above. Not every rock contains fossils. You will have to do considerable searching. The fossils seem to occur in clusters. You may find good fossil representation in small cabinet sized rocks or you may have to trim larger rocks down to manageable size. The matrix here is not terribly hard and much of it is easily split along existing fracture lines. With hammer and chisel, and skill and determination, you should be able to extract specimens from the large fossil bearing boulders.

Be particularly mindful of good rockhound etiquette here. Go about your collecting in a prudent, responsible manner. This is a high use area for environmentalists and conservation minded people. This site is very close to the Kanab Creek Wilderness Area and Grand Canyon National Park boundaries. So, disturb as little as possible, abide by the rules governing your right to collect fossils on public land, restore your collecting site to its natural state, and if you go prospecting, know exactly where you are.

*G.P.S. coordinates taken at the intersection of FDRs 423 and 642.

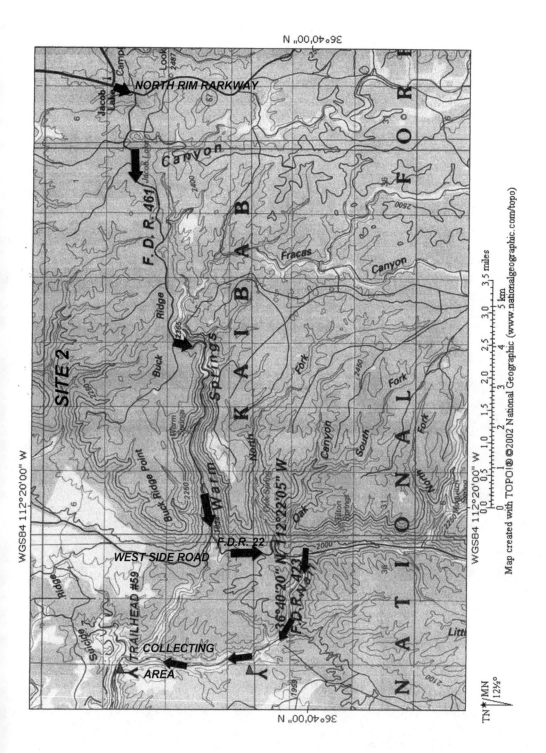

SITE 3

Crinoidea in Limstone Canyon

Difficulty Scale: 3 – 4 – 4   Seasons: Fall, Winter, Spring
Global Positioning System Coordinates: 34° 59' 12" N, 112° 27' 15" W*
Permian-Pennsylvanian Supai Group Limestone and Hermit Shale
U.S. Geological Survey 7.5 Minute Topographical Map: Paulden

The kiln in Limestone Canyon.

YOU CAN APPROACH THIS SITE FROM THE NORTH OR SOUTH. If you are coming from Ash fork (north), go south 18.5 miles on State Route 89 and turn right (west) onto Forest Road 573 which leads into Limestone Canyon. The turnoff is 1.5 miles south of the Hell Canyon bridge. If you are coming from Prescott (south), drive northward 31.4 miles from the intersection of State Routes 69 and 89 and turn left (west) onto F.R. 573. Follow F.R. 573 3.2 miles and park at the turnout beside the old kiln.

Best stay away from the old kiln as it is in serious disrepair. It is defying the law of gravity looking as if it will come crashing to the ground any second.

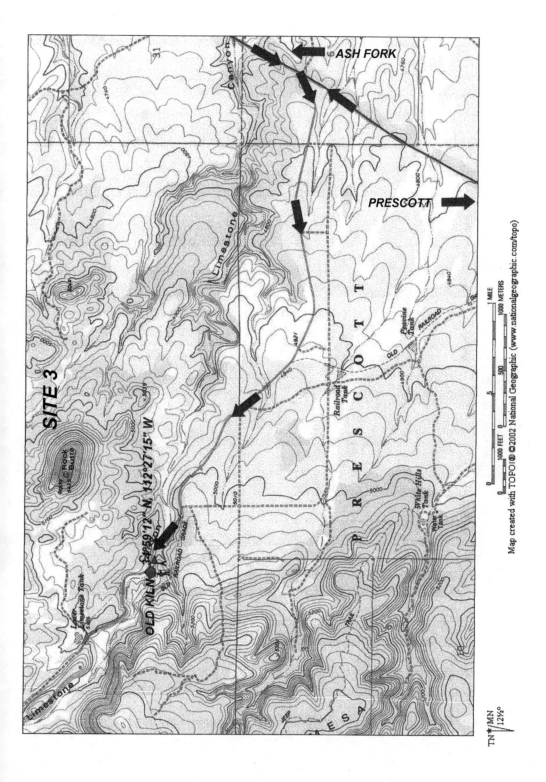

Compared to this sorry structure, the Leaning Tower of Pizza looks like a marvel of engineering.

On the hillside above the kiln are two quarries and above the quarries is an abandoned railroad grade. The strata here are Permian-Pennsylvanian Supai Limestone Formation . The lighter colored white and orange-red limestone contain crinoid hash. The battleship gray limestone which is often fossiliferous at other locations is barren here. The fossil bearing strata appears intermittently in the quarries, along the railroad grade, and on the slopes immediately east of the kiln. There are some small, loose fossil studded rocks scattered about the area. But, the best specimens are contained on top of the ledges exposed on the hillsides and along the railroad grade. To harvest these specimens, you will need a flat chisel and a heavy hammer to chip off fossil bearing plates.

G.P.S. coordinates taken at the parking area.

SITE 21

Plant Fossils at D K Well

Difficulty Scale: 5 – 4 – 5   Seasons: Fall, Winter, Spring
Global Positioning System: 43° 48' 45" N, 112° 00' 48" W*
Permian-Pennsylvanian Sedimentary Supai Limestone Group Formation
U.S. Geological Survey 7.5 Minute Topographical map: Clarkdale

FROM STATE ROUTE 260 IN CLARKDALE, turn north on the road leading to Tuzigoot National Monument. Drive .5 mile across the Verde River and turn left (west) on Sycamore Creek Road. Follow Sycamore Creek Road northward 5.9 miles to Buckboard Road. Turn right (east) on Buckboard and drive 1.5 miles to Road 9518 on your right (south). Turn right on Road 9518, go .3 mile, turn left (east), and drive another .3 mile to D K Well. Park at the well and hike from there.

The Verde Valley is composed of Pliocene–middle Miocene sedimentary limestone, mudstone, and sandstone. Much of it is fossiliferous although actual

View of D K Well from the road.

fossil occurrences are widely scattered over a large area. Reports of fossil discoveries in the Verde Valley area have circulated within the rockhound community for years. Most of these stories, however, have been vague as to location and inexact as to type of fossil. Serious fossil hunters may find this area ripe for exploration. The D K Well location may be typical of many other locations in the area awaiting discovery.

You can park at the well and explore the surrounding area on foot. The fossils here are flora hash (small stick and twig litter) and what appears to be molds or impressions of plant stems encased in limestone. Since the fossil bearing rocks are few and far between, you will have to roam over a wide area in search of a few good specimens. Upstream in the wash east of the well, there are a few large, dark colored boulders containing apparent plant stem molds. You will need a heavy hammer to break these boulders into specimen sized pieces. Downstream from the well, in the west wall of the wash, there are plant stem impressions exposed in the limestone stratum. It would take skill and patience to extract these specimens intact. The easiest specimens to collect, but also the hardest to find, are 6 – 12 inch long pieces of light colored, fossil bearing, limestone float.

*G.P.S. coordinates taken at D K Well.

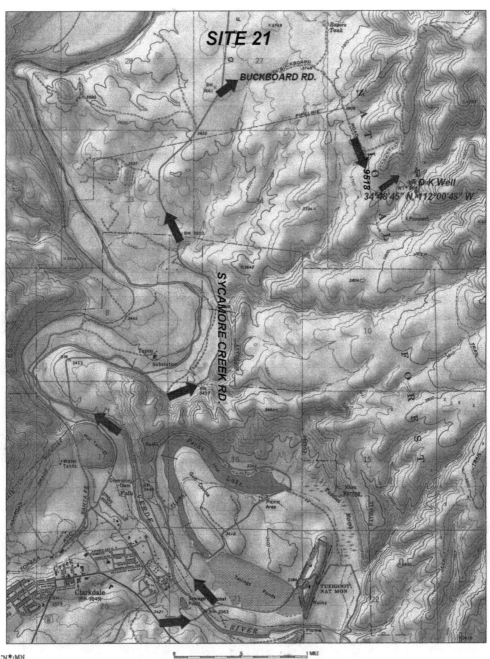

SITE 22

Marine Fossils West of Jerome

Difficulty Scale: 1 – 3 – 5   Seasons: Fall, Winter, Spring

Global Positioning System Coordinates: 34° 45' 34"N 112° 08' 05"W*

Paleozoic Sedimentary Martin and Redwall Limestone Formation

U. S. Geographical Survey 7.5 Minute Topographical Map: Munds Draw

FOLLOW U.S. ROUTE 89A to the point in Jerome where it makes a hairpin turn in front of the fire station. Right beside the firehouse, turn west onto Forest road 318, also known as Perkinsville Road. There is no road sign and Perkinsville Road looks more like an alley than a road next to the north side of the firehouse. This road is maintained for passenger car use even though it is rather rough and bumpy. The distance from the firehouse to the quarry turnoff is 1.6 miles. The turnoff to the quarry is at the crest of the hill on your right overlooking the United Verde Mine dumps and the town of Jerome. Turn in and park in the open area beside the footpath leading into the quarry.

    The collecting area is the quarry and the hill sides surrounding it. The quarry floor is flat and open, enclosed on three sides by cliffs 30 – 50 feet high. Large quantities of fossiliferous Redwall and Martin Limestone Formation rubble have fallen, either by natural erosion or human activity, from the cliffs above forming slides and spilling out onto the quarry floor. Fossil bearing rocks from several feet to just a few inches across are piled up several feet deep. Serious fossil hunters could spend weeks here systematically high grading these rubble piles for choice specimens. A word of caution, however. The cliff faces are extremely fractured, unstable, and undercut. So, watch out for falling rocks. Dislodging even one small rock from these cliffs could cause a fatal landslide. The hill sides surrounding the quarry are fairly easy to explore. The terrain is moderately sloped and not heavily vegetated. Although, there are some prickly pear cacti to avoid.

    Unlike many fossil collecting locations where specimens are sparse and widely dispersed, this site is (was) crawling (swimming) with ancient sea creatures. It may be wise to tour the collecting area first to familiarize yourself with its contents. Then, you can concentrate your efforts according to your

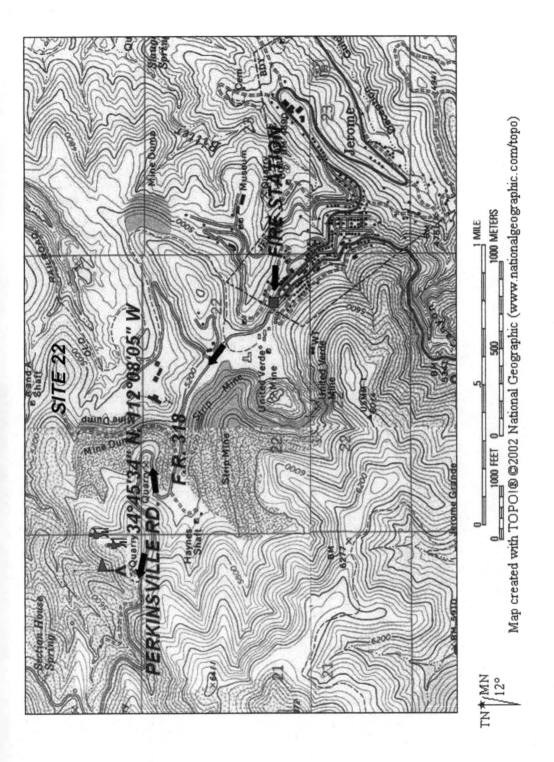

needs and abilities. Observe that two types of limestone predominate here. The first exposed stratum is a dull yellow-greenish, resinous looking limestone laced with calcite veins and crystal druses. The second stratum, which overlays the first, is an earthy, medium-gray color. It is solid and has a more uniform texture than the first stratum. Both strata contain fossils, although the second stratum contains significantly more. In the rubble piles in the quarry, dull-gray rocks of all sizes covered on all sides with crinoid hash are scattered among the calcitic material. Among the small crinoid debris, look for larger bivalves and gastropods. Nautiloids, ammonites, and trilobites have also been reported here, but in much smaller numbers. The ground surrounding the quarry is composed of exposed layers, ridges, and broken pieces of the dull-grey limestone stratum. Often, excellent specimens are visible imbedded in the surface of fairly large thick exposed rocks. Fortunately, this rock trims up in a fairly predictable manner. With a 2 – 3 pound hammer, a wide bladed chisel, and a little patience, you can usually split the large rocks and chip access material away from the fossil leaving it intact on a much smaller piece of matrix.

The turnoff to the fossil quarry.

G.P.S. coordinates taken at quarry entrance.

SITE 23

Crinoidea off State Route 89A

Difficulty Scale: 4 – 6 – 6   Seasons: Spring, Summer, Fall
Global Positioning System Coordinates: 34° 43' 46.2"N, 112° 08' 54.5"W*
Geology: Paleozoic Sedimentary Martin and Redwall Limestone Formation
U.S. Geological Survey 7.5 Minute Topographical Map: Hickey Mountain

GO SOUTH 5.1 MILES ON STATE ROUTE 89A from the Jerome fire station to a narrow dirt road on your right (west). If you are coming north from Prescott, the turnoff is on your left (west) 2.52 miles from the entrance to the Mingus Mountain Recreation Area. The turnoff road is at the south end of the guard rail on a sharp curve. Drive about 200 yards down the dirt road and park opposite the

The limestone quarry collecting site.

limestone quarry collecting site. You will have to pick your way through the rocks and vegetation to get up into the quarry on your right (north). The going is steep and rough.

The fossils here are small bivalves, crinoid parts, and sundry bits and pieces of marine life typical of the limestone roadcuts at the higher elevations of S. R. 89A. This little quarry is a convenient location for collecting the fossils common to this area. Although much limestone is exposed along the S.R. 89A roadside, collecting along this narrow, winding road is too dangerous. In addition to this quarry, there are several other dirt roads and turnouts along S.R. 89A to prospect as well.

The strata in the area is Martin and Redwall Limestone Formation. The fossiliferous limestone at this location is the darker, grayer stone. Much of it is in the form of large boulders. These boulders are very dense and hard to crack. Look for smaller pieces with fossil remains sprinkled across the surface. You may need a sledge hammer to split these rocks. Then, you may be able to trim them with your rock pick. The lighter colored limestone contains multi-colored calcite veins and vugs that are worthy of collecting.

*G.P.S. coordinates taken at quarry entrance.

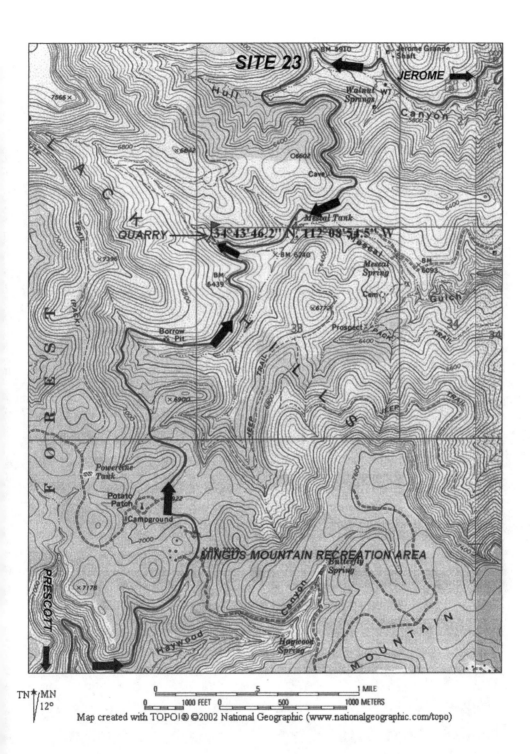

SITE 28

Colonial Coral near Chasm Creek

Difficulty Scale: 3 – 4 – 2   Seasons: Fall, Winter, Spring

Global Positioning System Coordinates: 34° 27' 07.0" N, 111° 49' 20.2" W*

Geology: Paleozoic Sedimentary Sandstone, Limestone, Dolomite.

U.S. Geological Survey 7.5 Minute Topographical Map: Horner Mtn

FOLLOW THE DIRECTIONS TO THE INTERSECTION OF Oasis and Salt Mine Roads in Verde Valley (see Site 27, page 111). Bear right (south) on Salt Mine Road and follow it 7.3 miles to a fork where the pavement ends. Go strait ahead onto unpaved Forest Road 574. Drive 3.3 miles on F.R. 574 to a turnout on the left (north) side of the road. Park here.

The collecting area is on the south side of the road along the base of the mountain range. Be sure to wear long pants and sleeves because the collecting

View of the collecting area across the road from the parking area.

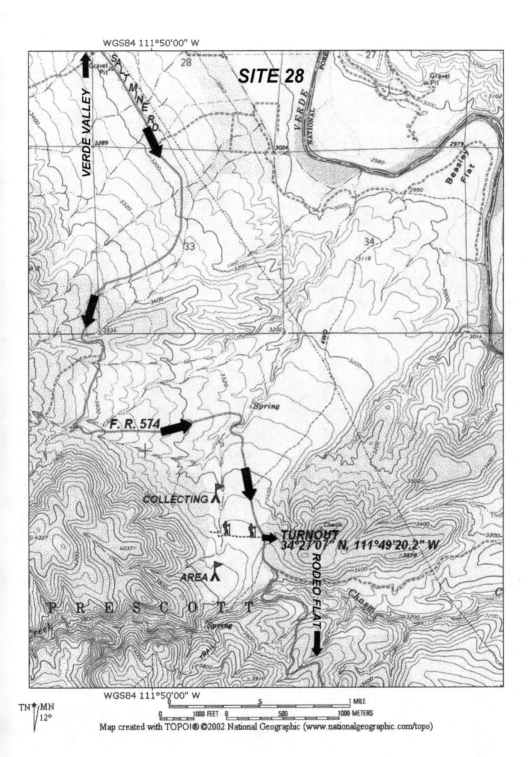

area and the route to it are heavily vegetated with prickly pear cactus and sticker bushes. After hiking about a quarter of a mile across a flat brushy plane, you will come to a barbed wire fence. The collecting area is between the fence and the mountains. Climb the fence. From here on, the terrain gets rougher and the vegetation gets thicker. Immediately beyond the fence is a deep wash. Once you find your way across the wash, you will encounter a more hilly landscape with several gullies running down out of the mountains.

There is not an abundance of fossil material here. But what there is is quite good. You will have to search over a wide area up and down, in and out of the hills and gullies. Look for softball size pieces of the darker gray limestone prominently studded with pieces of broken branched colonial coral limbs on the surface. Search carefully because there is a lot of similar looking limestone covered with plain chert and chalcedony bumps and ridges. The fossiliferous limestone pieces are widely scattered among the more common chert bearing limestone rubble.

*G.P.S. coordinates are at the parking turnout.

SITE 48

Mollusca at Battleground Ridge

Difficulty Scale: 3 – 4 – 6   Seasons: Spring, Summer, Fall

Global Positioning System Coordinates: 34° 32' 55" N, 111° 11' 32" W*

Geology: Permian Sedimentary Cherty Kaibab &
Toroweap Limestone Formation

U.S. Geological Survey 7.5 Minute Topographical Map:
Blue Ridge Reservoir

FOLLOW THE DIRECTIONS TO FOREST ROAD 123 in Site 49, page 153. From the intersection of Forest Roads 300 (Rim Road) and 123, turn left (north) on F.R. 123 and drive 7.6 miles past Site 49 to a fork. The left fork goes to a pumping station. Take the right fork and follow it 1.1 miles and park.

The collecting area is the forest floor within a radius of about 300 yards of where you park. There is not an abundance of fossils here. But, if you search carefully, you can retrieve some interesting specimens. On the south side of the

The parking area.

road where you park, look for a very large, old growth ponderosa pine (see the photograph on page 287). In front of it is a small pile of limestone rocks. Begin your search here. Among the predominant light grey, fine grained Toroweap Formation rocks, look for float from a different formation that is slightly darker and earthier colored. This rock is a burnt umber color, soft, porous, large grained, and looks more like mudstone than limestone. It is easily recognizable because its most distinguishing characteristic is its high concentration of fossil material. These rocks look like marine hash loosely cemented together with adobe. When cracked open, you will find both molds and casts of brachiopods, mollusks, crinoids, gastropods, and what appears to be sea urchin spines inside. You can break open the larger rocks rather easily with your rock pick. Work carefully so as to preserve the best specimens. On the slope to the left (south) of the big tree, look for small pieces on the ground. At the top of the slope near a large campfire pit, there is a fossiliferous vein of this material about 15 feet long running in a north-south direction.

G.P.S. Coordinates taken parking area.

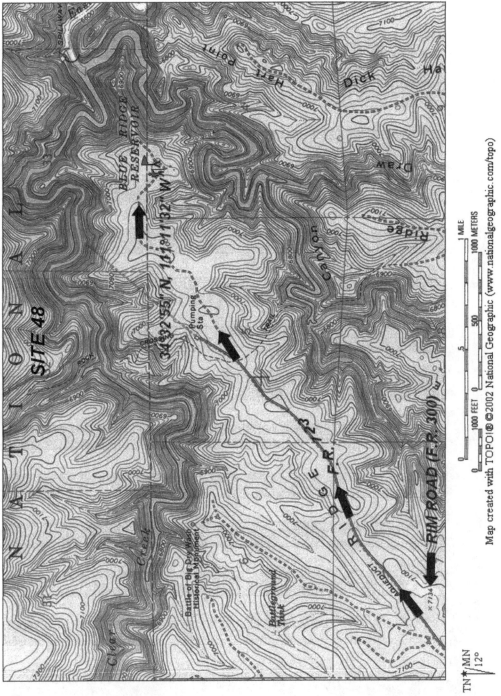

SITE 50

Crinoidea on Forest Road 29A

Difficulty Scale: 6 – 4 – 5   Seasons: Spring, Summer, Fall

Global Positioning System Coordinates: 34° 20' 41.8" N, 111° 06' 54.7" W*

Geology: Carboniferous Sedimentary Redwall and Naco Limestone Formations

U.S. Geological Survey 7.5 Minute Topographical Map: Promontory Butte

FROM THE INTERSECTION OF STATE ROUTES 87 and 260 in Payson, go east on Route 260 16.7 miles and turn left (north) on the paved road leading to the Tonto Creek Recreation Area and the fish hatchery. The road is about .2 mile east of Kohls Ranch. Follow this road 2.2 miles and turn left (west) on Forest Road 29, also known as Roberts Mesa Road. After driving 1.5 miles, on F.R. 29, you will see F.R. 29A, a rough un-maintained road, leading up a hill on your left (south). Turn left onto this road. The collecting area is between .7 and .8 mile down F.R. 29A on the uphill slope on your right (north) and in the washes.

Beginning of the crinoid collecting area on Forest Road 29A

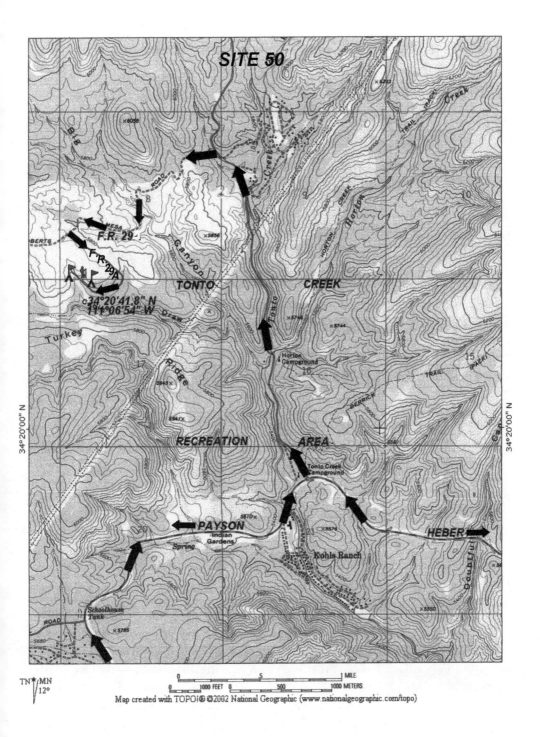

This location may well yield the best crinoid specimens in Arizona. Although crinoid is the predominant fossil here, brachiopods and bivalves appear in lesser numbers as well. Unlike most crinoid collecting sites where the stems and pieces are small and widely scattered across limestone surfaces, the specimens at this site are large and so tightly packed together that limestone slabs look as if they were composed entirely of crinoid parts. Fortunately, the Permian Naco Limestone Formation strata here tend to exfoliate in thin, 1 – 4 inch thick, slabs. Crinoid parts are liberally distributed more or less evenly throughout the thickness of the slabs. Typically, at most Arizona crinoid locations, the individual disks that form the crinoid stem have become detached from one another and lie randomly scatted about the sedimentary strata. However, here you can find unusually long strands of crinoid limbs ¼ – ¾ of an inch in diameter, 3 – 4 inches long, containing 20 – 30 disks. Some strands resemble a stack of washers that fell over in a straight line leaving each one separated from the other but, leaning on each other at a 45 degree angle. Search for plates displaying these characteristics as they are very showy and are highly collectable.

Most of the fossil remains here are imbedded in a light tan or light gray limestone. Surface specimens are exposed about 50 percent. They are about the same color as the limestone that surrounds them. Contained within the limestone, however, is a great abundance of chert patches and veins. Scattered across the road and the ground, you will see a large amount of large, colorful, chunky chert that has weathered out of the limestone strata. Chert is usually rather ordinary uninteresting stuff. But, some of the chert at this site is bright, colorful, well-patterned, and worthy of collecting for its lapidary potential. Some of it, however, is honey combed with open spaces and seams. This material contains the most remarkable and best quality fossil specimens found in this area for they have become chertized and often appear almost fully exposed as if they had been strategically placed on top of a layer of gemmy chert. Red, pink, yellow, and orange brachiopods and bivalves are sometimes sheltered in the cracks and crevices of cherty formations. They look as it they had been fashioned from colorful gem quality agate. These specimens are not common, but they are worth looking for.

Collecting in the pine forest here is fairly easy and at nearly 6000 feet elevation, it is very pleasant in the summertime. Search among the trees on the slopes on the right hand (north) side of the road. The forest is fairly open with few bushes or steep ledges to impede your progress. Look for open spaces where the ground is littered with broken pieces of limestone rocks. In the washes, examine the stratum that forms the bottom of the water course. In some places, that stratum is very fossiliferous and, when exposed, tends to slough off large fossil packed plates 2 – 3 feet across.

*G.P.S. coordinates taken at beginning of collecting area.

SITE 51

Porifera, Braciopodia, and Crinoidea along Forest Road 237

Difficulty Scale: 2 – 1 – 1   Seasons: Spring, Summer, Fall
Global Positioning System Coordinates: 34° 20'49.8" N, 110° 49.' 53.1" W*
Geology: Permian Sedimentary Kaibab and Toroweap Limestone Formations
U. S. Geological Survey 7.5 Minute Topographical Map: O W Point

FROM THE INTERSECTION OF STATE ROUTES 260 and 87 in Payson, turn east on State Route 260. Go 34.7 miles to Forest Road 237 between mileposts 286 and 287. You will pass Kohls Ranch, Christopher Creek, and the intersection of S.R. Route 288 along the way. If you are coming from the East on S.R. 260, F.R. 237 is 18.3 miles from the Mogollon Rim High School in Heber. Turn north on F.R. 237 and drive 1.5 miles to F.R. 237B. Turn right on F.R. 237B and follow it .1 mile to the clearing under the high wire transmission line. Park there and begin searching.

This is more of a collecting locality than it is a specific site. If you look

The wide swath cut through the forest under the power line on F.R. 237B.

north, you will see a wide swath cut through the forest under the power line. This is the easiest place to collect. In fact, you will find fossils right where you park. You will find brachiopods and crinoid stems on the surface of the rocks that litter the ground. You will also find brachiopods loose in the soil. By cracking these rocks open, you can reveal fossil sponges that are encased within. You can collect northward along the power line for several miles if you like. Be selective. Collect only the most complete and best preserved specimens. There is plenty of material here scattered over a wide area. So, there is no reason to hoard every fossil scrap you see. If you enjoy a walk in the woods, you can search for exposed fossil bearing Permian limestone outcrops among the trees. Sometimes, you can discover white brachiopods poking up through the pine needles like mushrooms. In this area, Forest Service rules allow vehicle traffic on numbered roads only. But if you bring your ATV, you will be able to explore a large network of ATV only roads that wind throughout the forest in this locality. Bouncing through the forest on your ATV searching for fossils is great fun.

The predominant fossils here, as on most of the Magellan Rim, are brachiopods and crinoid stems. Unfortunately, most fossils found here are incomplete and imperfectly preserved. The brachiopod molds and casts are usually of one shell only. Pairs of two hinged opposing valves are rare but not impossible to find. Some have completely weathered out of the Permian limestone strata and others are partially exposed. The brachiopods here are 1 – 2 inches across and spherically shaped. Most are not particularly well preserved, but if you search hard enough, you will likely uncover a few fairly presentable specimens. Likewise, you will not find a complete crinoid. Rather, you will find jumbles of disarticulated fossilized arms, columns, heads, and other animal parts that apparently were randomly scattered as the animals became dismembered. Most preserved parts are long and spindly or round in shape. Several dozen often are perched together in clusters on rocks a foot or more in length.

Occasionally, you may be fortunate enough to encounter a crinoid jumble resting on top of or beside a bryozoan fossil. Look for a repetitive pattern across the face of the rock looking as if a piece of lacy loosely woven fabric or net had left an impression on a muddy sediment.

G.P.S. coordinates taken in the clearing under transmission line.

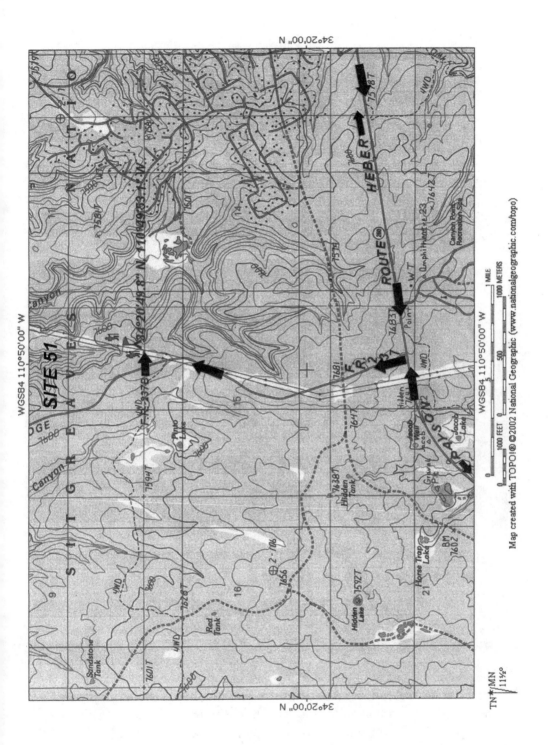

SITE 54

Colonial Coral
South of Control Road

Difficulty Scale: 6 – 4 – 4   Seasons: Spring, Summer, Fall

Global Positioning System Coordinates: 34° 22' 39" N, 111° 16' 23" W*

Geology: Early Paleozoic Gray Limestone and Dolomite
and Red-Brown Sandstone

U.S. Geological Survey 7.5 Minute Topographical Map: Kehl Ridge

The rough, steep road leading south from Control Road to the collecting area.

THE NEXT THREE COLLECTING SITES, 54 – 56, are located along Control Road between State Routes 87 and 260 below the Mogollon Rim north-east of Payson. Sites 54 – 56 are merely examples of many known and yet to be discovered fossil locations that abound along the Mogollon Rim and in the Tonto and Coconino National Forests. To reach site 54, go to the intersection of State Routes 87 and 260 in

Payson. Drive north on S.R. 87 2 miles, turn right (north-east) on Houston Mesa road, and follow it 10 miles to the intersection of Houston Mesa and Control Roads. Turn right (east) on Control Road and drive .9 mile where you will come to a rough, steep road on your right (south). The first 50 feet are difficult. Follow this road .3 mile and park at the bottom of the steep, eroded 4-wheel drive hill. The collecting area is the ridge line at the top of the hill. If you are coming from the north down S.R. 87, turn left (east) onto Control Road and drive 10.2 miles to the turnoff to the collecting area. As you proceed up the road to the fossil collecting area, you will see thousands of 1–2 inch sized geodes scattered across the road and over the forest floor. Unfortunately, the reason that there are so many is that they are ugly and not worth collecting.

The hike up the road to the top of the ridge is short but steep. The collecting area covers a good amount of territory, stretching along the top of the ridge in both directions and down both the north and south slopes. When you reach the top, you will be greeted by thousands of branching and honeycomb-like colonial coral fossil specimens protruding from an exposed, weathering, limestone stratum that covers the top of the ridge. You can attempt to retrieve large fossil bearing plates with hammers, chisels, and crow bars, or you can search in the soil for smaller more manageable specimens. You can find small pieces of limestone bristling with well exposed and well preserved coral branches and colonies. On the south slope of the ridge, roads lead off to the east and west to additional fossil outcrops. On the north side of the ridge, small mollusk fossils occur along a faint, overgrown road that leads west from the parking area.

*G.P.S. coordinates taken at the parking area.

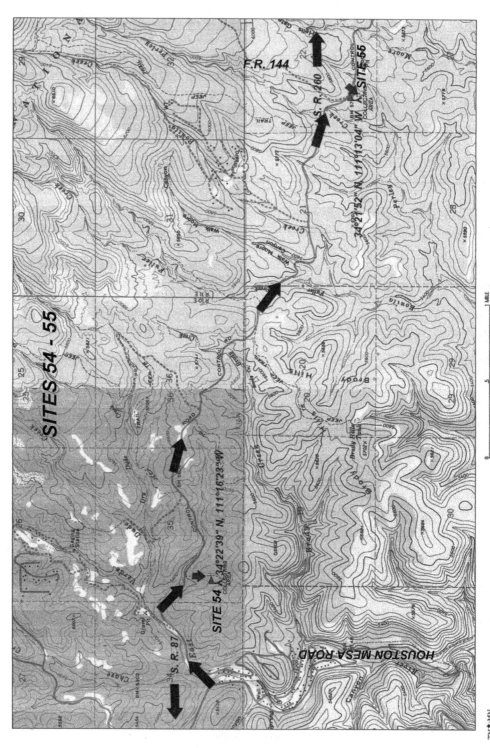

SITE 55

Coral Fossils
South of Forest Road 144

Difficulty Scale: 3 – 4 – 4   Seasons: Spring, Summer, Fall

Global Positioning System Coordinates: 34° 21' 52" N, 111° 13' 04" W*

Geology: Early Paleozoic Gray limestone and Dolomite
and Red-Brown Sandstone

U.S. Geological Survey 7.5 Minute Topographical Map: Diamond Point

TO REACH THIS SITE, follow the directions to Site 54 on page 296. From the turnoff to Site 54, continue driving east on Control Road 3.8 miles where you will come to F.R. 144 on your left (north). The collecting area is .1 mile into the forest on your right (south) opposite F.R. 144. You can drive in .1 mile on a faint, little, cul-de-sac road that ends on top of a gentle rise. The collecting area is the top of a low round hill, where you can park, and the slopes that descend from it to the west and north.

The entrance to the collecting area from Control Road opposite Forest Road 144.

There are no more easy pickings here since this location has been heavily collected for years. However, if you search diligently and carefully, you can discover some excellent coral and brachiopod specimens in the soil. Be sure to wear long pants and sleeves to protect yourself from the patches of vicious pricker bushes that grow on the hill sides. The fossil matrix here is an eroding, dark, red chert that is scattered across the hill sides and gullies. The prized specimens here are coral heads, some up to 8 inches across, and solitary, cornucopia shaped, Rugosa or "horn" corals with openings up to 1½ inches across. You can also find good brachiopod and crinoid molds and casts here. Although all the fossils here are chertized, some brachiopod molds still contain partial shell remnants.

*G.P.S. coordinates taken at the top of the hill.

SITE 56

Crinoidea off Forest Road 29

Difficulty Scale: 2 – 3 – 6   Seasons: Spring, Summer, Fall
Global Positioning System Coordinates: 34° 19' 57" N, 111° 10' 40" W*
Geology: Early Paleozoic Gray Limestone and Red-Brown Sandstone
U.S. Geological Survey 7.5 Minute Topographical Map: Diamond Point

View of the gate at the entrance to the quarry.

IF YOU ARE APPROACHING THIS SITE from State Route 260, turn west onto Control Road, drive 4.1 miles to Forest Road 29, also known as Mead Ranch and Roberts Mesa Roads, and turn right (north). If you are traveling east on Control Road from sites 54 and 55, drive 4.2 miles from Site 55 and turn left (north) onto F.R. 29. Go .3 mile and turn right (east) onto the road leading into the quarry. You will have to park in front of and walk a short distance beyond the locked gate

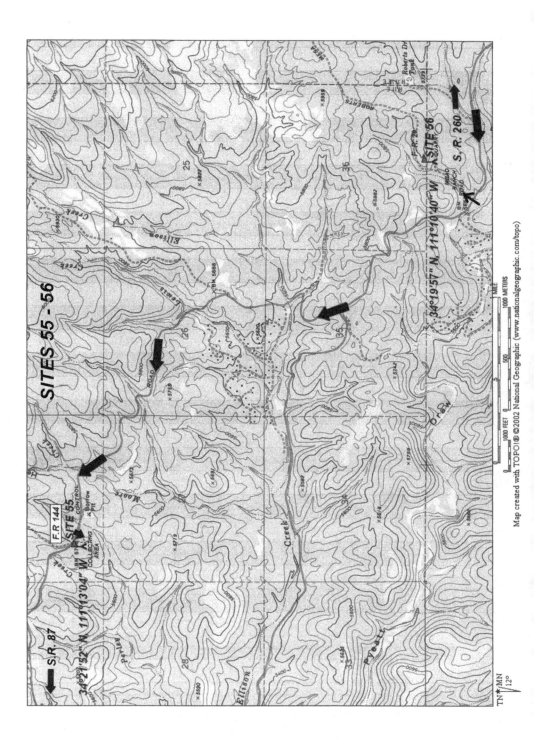

where you will come to the collecting area on the slope on your right immediately behind the gate.

Getting to this site is the easy part. Most of the fossils here are contained in rather large rocks beside the road at the bottom of a hillside. You can harvest specimens from the surface of these rocks by chipping pieces off with a heavy hammer and chisel. If you would rather not work that hard, you can search for smaller pieces in the dirt farther up the graded slope. The crinoids here are rather small, measuring 1/8 – 1/4 inch across.

*G.P.S. coordinates taken at the quarry entrance.

SITE 59

Rugosa off Forest Road 138

Difficulty Scale: 2 – 4 – 5   Seasons: Fall, Winter, Spring

Global Positioning System Coordinates: 31° 41' 21" N, 110° 57' 30" W*

Geology: Cretaceous-Late Jurassic Sedimentary Bisbee Group Limestone

U.S. Geological Survey 7.5 Minute Topographical Map:
Amado, Mount Hopkins

The little, narrow road leading from Forest Road 138 up to the collecting area.

FROM THE INTERSECTION OF INTERSTATES 10 and 19 in South Tucson, go south on I-19 34.5 miles and turn off at exit 48. After exiting, turn left (east), go under the overpass, and turn left (north) onto the frontage road. Follow the frontage road north for 1.7 miles and turn right (east) on Elephant Head Road. Drive 1.5 miles on Elephant Head and turn right (south-east) on Mt. Hopkins Road. Because this road is fairly new, some maps do not show it. Go 5.5 miles on Mt. Hopkins Road, turn left (east) onto Forest Road 138 and drive .8 miles to a 90° bend in

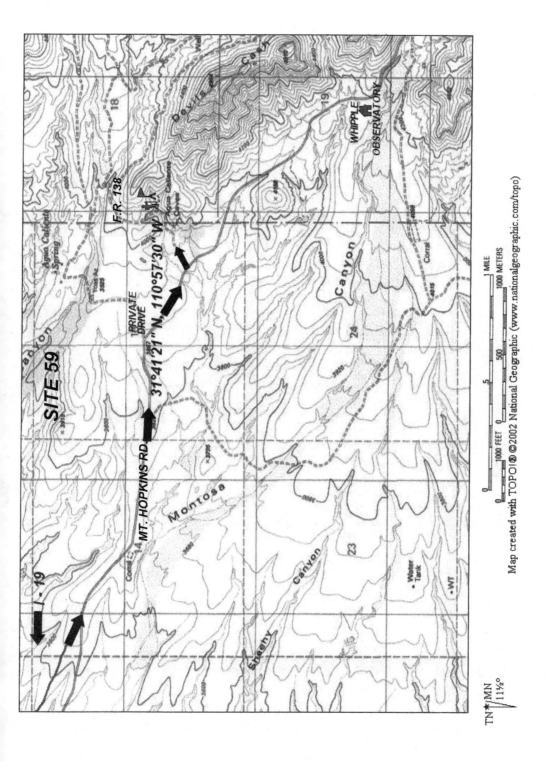

the road where you will see a little, narrow road leading into a small canyon on your right. From here, you will need to park and hike up the canyon to the saddle at the top of the ridge. Within about 200 feet, the narrow road comes to an end and becomes a footpath. Follow the path 400 – 500 feet up to the saddle. The collecting area begins at the saddle and continues across the top of the ridge.

There is not an abundance of fossils at this location. But, there are some choice solitary coral specimens protruding from the Bisbee Limestone Formation. Search carefully for small fossil bearing pieces in the rubble that is scattered across the hill side and hiding under the vegetation. Along the top of the hill, you can see solitary coral specimens protruding from the exposed, weathered limestone. With a heavy hammer and chisels, you may be able to liberate suitable specimens intact. The Bisbee Formation continues along the hillsides on the north side of Mt. Hopkins Road about as far as the Smithsonian building that services the Whipple Observatory atop Mt. Hopkins. While you are here, the observatory information center is worth a visit.

G.P.S. coordinates taken at the parking area on F.R. 138.

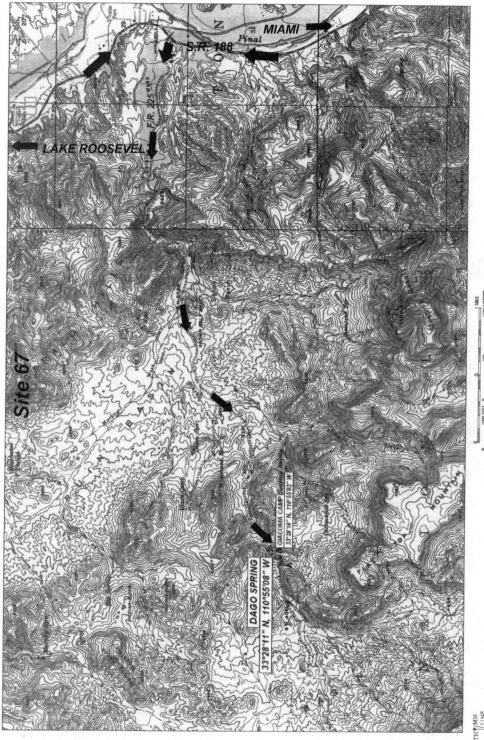

SITE 67

Trace Fossils at Dago Spring

Difficulty Scale: 4 – 7 – 6   Seasons: Fall, Winter, Spring
Global Positioning System Coordinates: 33° 28' 11" N, 110° 55' 08" W*
Geology: Paleozoic Sedimentary Limestone, Shale, Dolomite
U.S. Geological Survey 7.5 minute Topographical Map: Inspiration

From the intersection of U.S. Route 60 and State Route 188 in Miami, drive 5 miles north on S.R. 188 to mile marker 220. Turn left (west) .15 mile beyond mile marker 220 onto Forest Road 225. The road leads downhill. The F.R. 225 road marker is at the bottom of the hill. Older maps including the Inspiration 7.5 minute topographical map will show the old narrow, winding S.R. 88 rather than the new, improved S.R. 188 that is being built to replace it. Both roads run roughly parallel to each other. Follow F.R. 225 4.4 miles to Dago Spring. Except for a few shallow rain gullies, F.R. 225 is almost good enough for passenger car

The gate across the road, walnut trees, and windmill at Dago Spring.

travel. When you arrive at Dago Spring, you will encounter a large iron gate across the road and you will see a corral with a large windmill and water cistern in a grove of walnut trees on the left (south) side of the road.

The collecting area is near the top of the steep hill immediately beyond the fence and the little wash that runs parallel to the south side of the road. Climb the fence and walk 100 feet or so to the wash. Here, you will be face to face with a steep tailing slide. At the top of the slide is a cliff that once contained fossils, but the supply here is now pretty much exhausted. Rather than trying to climb up the tailings slide, turn to your left (east) and follow the narrow footpath that leads out of the wash and up the hill through the bushes. This trail is rough, very steep, and overgrown with sticker bushes. Wear long sleeves and sturdy climbing boots. The trail leads upward and to the left (eastward). After scrambling up through the thick bushes for 100 feet or so, you will begin to see tailings from the collecting area above. Then, you will come to a vertical cliff 8 – 10 feet high. Above the cliff, there are several shallow prospect holes. The stratum containing the trace fossils is 1 – 2 feet from the bottom of the cliff. Look for the lightest colored layer, 1 – 4 inches thick.

Trace fossils are evidence of a prehistoric animal's activity rather than the fossilized remains of the actual animal itself. The trace fossils at this site are pre-Cambrian worm burrows (*Tigillites bohmei*) contained in Mescal Limestone Formation. They are not nearly as exciting as T-rex tracks, but they provide an interesting dimension to a fossil collection for those who are interested. To harvest these trace fossils, carefully extract pieces of the fossil bearing stratum with a small chisel or the point of your rock pick. The stratum is fractured vertically in flat sheets allowing you to remove pieces fairly easily. Look for little round circles 1/16 – 1/4 inch in diameter. These are worm burrows that have filled in with silica. They are colored in shades of white and brown. Some have a brown circle within a larger white circle like a bull's eye. Depending on how the stratum fractures, some will protrude above the face of the rock.

In addition to the worm burrows, there are two other types of rocks farther up the hill to collect. The first is in the shallow excavations just above the fossil bearing cliff. Coursing through the limestone strata are multi-colored bands of silica. Some are transparent like agate and others are more opaque like jasper. Colors are red, orange, brown, and green pastels. The second is green, gemmy serpentine that occurs intermittently on the slopes between the cliff and the top of the hill.

There is yet another collecting location on the south side of the road about .2 miles before reaching Dago Spring. This is Giacoma Camp. Although this was a copper mine, there is no evidence of copper minerals. Look for a steep rusty colored cliff and tailing slide east of Dago Spring. The tailing slide contains pieces of limestone stained and banded in picturesque patterns.

*G.P.S. Coordinates Taken at the iron gate.

SITE 69

Bivalva on State Route 77

Difficulty Scale: 1 – 3 – 6   Seasons: Fall, Winter, Spring

Global Positioning System Coordinates: 33° 00' 04" N, 110° 45' 59" W*

Geology: Paleozoic Sedimentary Martin and Redwall Limestone Formation

U. S. Geological Survey 7.5 Minute Topographical Map: Hayden

OF ALL OF THE COLLECTING SITES IN THIS BOOK, this one is the easiest to find. Just go to the corner of State Routes 177 and 77 in Winkelman. From there, go northeast 1 mile on S.R. 77 to the quarry on the left (north-west) side of the road. There is room for one or two vehicles to park at the entrance to the quarry behind the guardrail. There is also a turnout overlooking the river on the northeast side of the road about 200 yards beyond the quarry where several vehicles can safely park.

View of the fossil quarry from State Route 77.

State Route 77 between Winkelman and Globe cuts through a variety of Cenozoic and Mesozoic limestone and mudstone deposits. Although, some of the strata along this road is known to be fossiliferous, specimens are widely dispersed throughout the limestone rock. In any case, there are several road cuts, bankings, and slopes along this road that might surrender specimens if diligently prospected. While not plentiful, the Winkelman quarry contains the greatest concentration of fossils discovered so far along S.R. 77.

The strata in the quarry is Pliocene-middle Miocene limestone basin deposits. Although the strata in the quarry is horizontal, the strata in the cliffs and slopes along S.R. 77 is dramatically buckled and tilted as much as 45° or more as a result of Tertiary faulting. Look for two types of fossiliferous rock on the quarry floor. The first is a light gray rounded rock covered with small sized, tightly packed crinoid hash. This type of rock is randomly scattered about the quarry floor. The best way to harvest specimens from these rocks is to use a hammer and flat chisel to break off fossil bearing chips from the face of the rock. The second type of fossiliferous rock is a light colored stratum that lay exposed in a ledge 2 – 3 feet high across the middle of the quarry floor. The rock is packed with brachiopod, mollusk, and crinoid hash. Since this stratum tends to exfoliate horizontally, the best way to extract specimens is to apply a wide, flat chisel on the edge of the ledge about 1 inch from the top and carefully tap it causing the stratum to split off in large sheets exposing the fossils within. The fossils here are usually tightly encased in limestone. To reveal the fossils more prominently and completely, you can soak the limestone slabs in a mild acid solution for a time to remove some of the limestone from around the fossil structures.

*G.P.S. coordinates taken in the quarry.

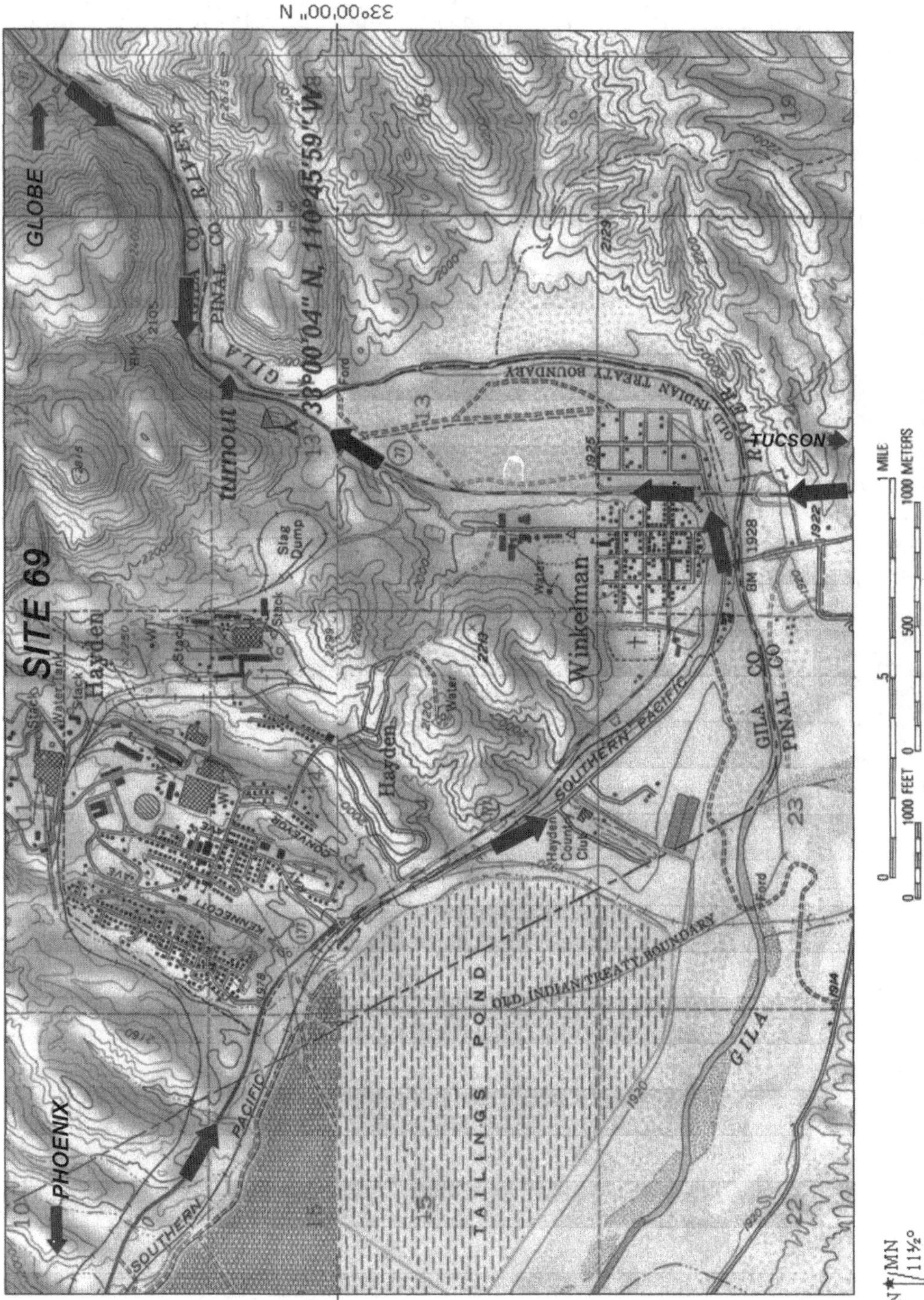

Map created with TOPO!® ©2002 National Geographic (www.nationalgeographic.com/topo)

SITE 73

Marine Fossils in French Joe Canyon

Difficulty Scale: 5 – 4 – 6   Seasons: Fall, Winter, Spring

Global Positioning System Coordinates: 31° 48' 29" N, 110° 23' 03.7" W*

Geology: Permian-Pennsylvanian Naco Group Limestone Formation

U.S. Geological Survey 7.5 Minute Topographical Maps:
McGrew Spring and Apache Peak

DRIVE SOUTH ON STATE ROUTE 90 11.3 miles from Interstate 10, exit 302, in Benson. Turn right (west) on the un-maintained dirt road that leads to French Joe Canyon. If you are traveling north from Sierra Vista, turn left (west) 8.4 miles from the intersection of State Routes 90 and 82. Follow the dirt road .8 mile to a fork and bear left (south-west). After driving 1.6 miles on this road, you will begin to enter French Joe Canyon where fossiliferous rocks begin to appear. The collecting area extends well into the canyon. Dry Canyon and Mine Canyon a few miles south of French Joe Canyon also contain fossils.

Beginning of the collecting area at the entrance of French Joe Canyon.

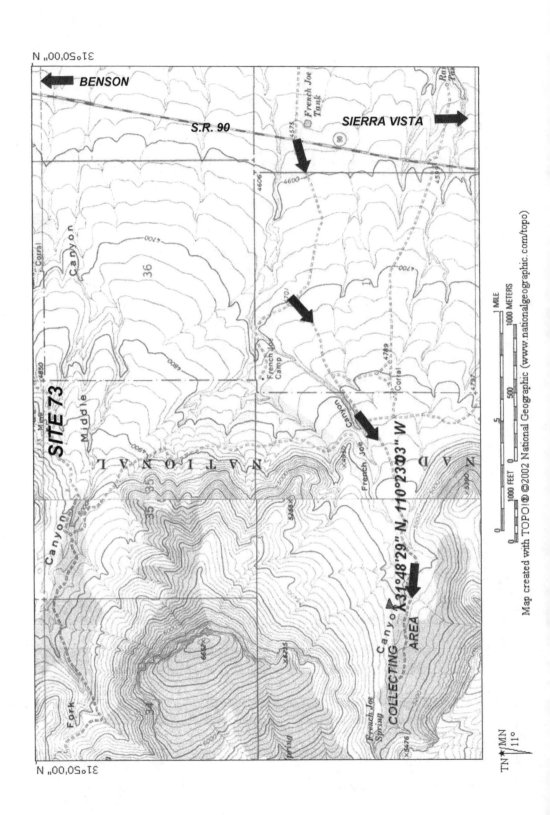

Look for the typical battleship gray, smooth, rounded Pennsylvanian and Permian limestone rocks sprinkled with crinoid, mollusk, and brachiopod hash. You may have to hike around a bit since the fossil bearing rocks are widely scattered over the landscape. Many of these rocks are large, one to three feet across, and need to be reduced to manageable size with a sledge hammer. The fossils seem to be confined to the surface of the rocks. If you are careful and lucky, you can chip off specimen plates by striking the edge of the fossil bearing rock face. Petrified coral fossils have also been reported here.

*G.P.S. coordinates taken on the road at the beginning of the collecting area.

SITE 76

Gastropoda in the Mule Mountains

Difficulty Scale: 3 – 3 – 5   Seasons: Fall, Winter, Spring

Global Positioning System Coordinates: 31° 26' 32" N, 110° 01' 39" W*

Geology: Permian-Pennsylvanian Sedimentary Naco Limestone Formation

U. S. Geological Survey 7.5 Minute Topographical Map: Hereford, Ariz.

FROM ROUTE 92, ABOUT HALF WAY BETWEEN Sierra Vista and Bisbee, turn north on S. Rio Vista Road which is 3.2 miles east of Hereford Road and 3.3 miles west of Wilson Road. Go 1.3 miles on S. Rio Vista and turn left (west) on W. Calle 5. Drive .5 mile on W. Calle 5, turn right (north) onto S. Via Liberacion, and proceed 2.3 miles to a fork. Bear left at the fork and continue 1.3 miles to a faint road on your left that leads to the low limestone hills which is the collecting destination. Drive about .2 mile and park. The walk to the base of the low hills is about 200 yards.

The faint road leading to the low limestone hills.

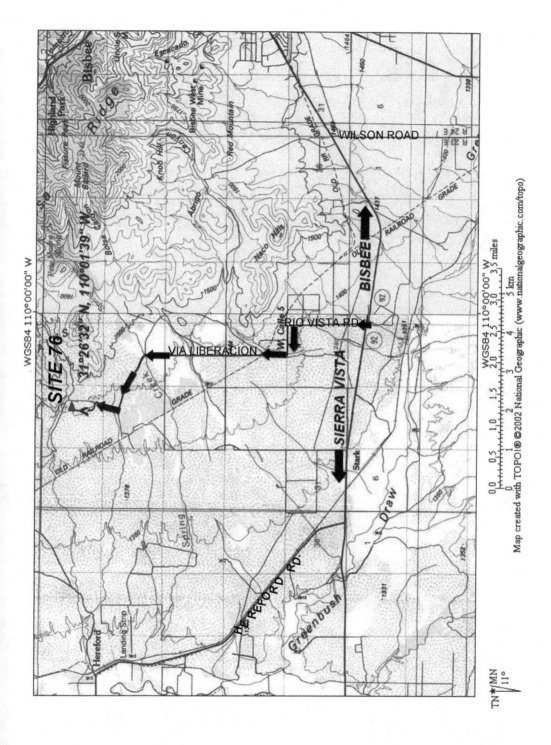

This location is just north of the Naco Hills which also contain marine fossils in the Naco formation. As with most Arizona fossiliferous strata, fossils here are not in abundance. You have to search diligently over a wide area and be satisfied with discovering a few good specimens. The easiest method is to search through the eroded rubble that litters the ground at the base of the hills. Look for small light colored ammonites, 1 to 2 inches across, in the light gray limestone. Search carefully because the ammonites are the same texture as the limestone and only a shade lighter in color. If you walk up the hillside, look for small crinoid remains, 1/16 to 1/4 of an inch across in clusters perched on top of weathered exfoliating limestone plates. The best way to harvest these is to get a flat chisel underneath the plates and carefully lift then free. If you dig, as others have before, be sure to fill in your holes.

*G.P.S. coordinates taken at the base of the low limestone hills.

SITE 89

Gastropoda at the Cochise Mine

Difficulty Scale: 5 – 5 – 4   Seasons: Fall, Winter, Spring
Global Positioning System: 31° 58' 59.0" N, 109° 10' 17" W*
Geology: Lower Permian-Pennsylvanian Sedimentary Horquilla Limestone
U.S. Geological Survey 7.5 Minute Topographical Map: Portal

View of the Cochise Mine from Foothills Road.

FROM INTERSTATE 10 AT SAN SIMON, take exit 382. Drive east about 1 mile on the south frontage road, Power Road, to the intersection of Power and Nolan Roads. Turn south on Nolan which is also known as the San Simon-Paradise Road. Follow Nolan Rd. southward 16.8 miles until you arrive at a fork. Bear left (east) at the fork onto Foothills Road which leads to Portal. The turnoff to the Cochise Mine is 2.25 miles down Foothills on the right (west). You can see the mine tailings piles back in the hills as you approach the turnoff. If you are coming from Portal, the distance is 6 miles from the intersection of Portal and Foothills Roads. The turnoff is a faint trail that goes through a broken down fence and off

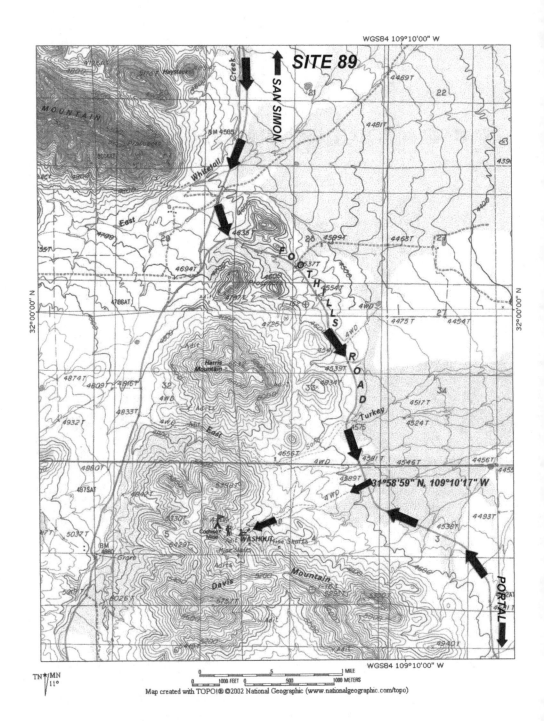

through the bushes on the west side of Foothills Road. You can drive almost all the way to the mine complex before the road washes out.

The collecting area is the slopes below the tailings piles. Here you will find small mine dumps containing some low grade azurite and malachite specimens. Most of these are not worth collecting. Scour the hillsides for Carboniferous and Devonian Horquilla Formation, a type of Naco Formation, gray and dark gray, almost black, limestone. The dark gray stratum contains ammonites and mollusca. Usually, these fossils are broken open revealing the curled edge of the shell filled with the same material as the matrix. The shell edge is white making a good contrast with the nearly black matrix. Unfortunately, no internal chambers are visible. The lighter gray strata contains small, 1/16 to 1/4 inch across, crinoid parts. Although, there is an ample amount of fossiliferous float covering the hillside, actual fossils are scarce. You will have to search patiently for a few good specimens. The white fossil shells fluoresce a dull red. Additionally, the second highest mine dump on the north side of the complex contains a limited amount of very attractive pale green and maroon striped rhyolite.

G.P.S. coordinates taken at the corner of the turnoff to the mine and Foothills Road.

SITE 90

Crinoidea at the Willie Rose Mine

Difficulty Scale: 5 – 4 – 6   Seasons: All

G.P.S. Coordinates: 32° 05' 46" N, 109° 15' 15" W*

Geology: Cretaceous-Late Jurassic Naco Group Gray Limestone

U.S. Geological Survey 7.5 Minute Topographical Map: Blue Mountain

View of the Willie Rose Mine from the road.

FROM INTERSTATE 10 IN SAN SIMON, take exit 382. Drive east about 1 mile on the south frontage road, Power Road, to the intersection of Power and Nolan Roads. Turn right (south) on Nolan, also known as the San Simon-Paradise Road. Drive south on Nolan Road 9.5 miles and turn right (west) onto a dirt road. Go .5 mile to a fork and turn left (south). Follow this road 1.4 miles where you will come to a second fork. Bear right (north-west), go 1.8 miles, and turn left (south). Follow this road 2.5 miles to the Willie Rose Mine. You will not be able to see the mine on the hillside until you are almost there. Look for an open space among the prickly pear cacti to park.

Along the extreme eastern edge of the Chiricahua Mountains between the Willie Rose Mine area and the town of Portal, strata of fossiliferous Naco Limestone Formation are briefly and intermittently exposed. The Willie Rose Mine is located in a cluster of low hills formed by a stratum of lower Permian Concha Formation Limestone, which is a type of Naco Formation. This formation is fossiliferous and cherty. The mine itself is rather ordinary. The dumps contain mediocre traces of copper minerals and some common red fluorescing calcite which may be mildly interesting to the collector. But, it is the fossils in the hillsides that are the main attraction here. The hills are composed of a series of low, battleship gray ledges that stair-step up the hillsides from bottom to top. Climbing up is rather difficult, however, because of the thick prickly pear forest that grows here. Protruding from the limestone ledges are numerous, large, gnarly pieces of chert that look like fossilized pieces of forest floor debris. Randomly scattered among this material are zones of marine hash containing brachiopod, mollusk, and crinoid remains. You may have to search over a wide area to find fossiliferous pieces small enough to carry. Most of the hillside is solid ledge. Otherwise, you will have to perform some hard labor with a heavy hammer and chisel to liberate specimens from the rocky ledges. It may be worth while to hike back into these hills and search this area more thoroughly for more fossil bearing strata.

*G.P.S. coordinates taken at the Willie Rose Mine.

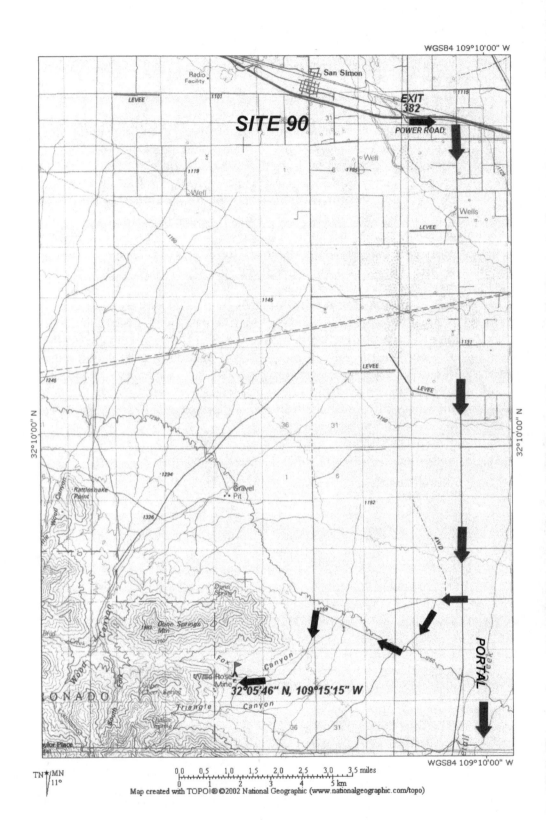

Fluorescents

OF THE 3,000 OR MORE MINERAL SPECIES in the world, collectors would regard most as merely rubble. The joy of collecting comes from rummaging through and discarding acres of rubble in search of a few worthy and aesthetically pleasing specimens. We are drawn to what we can see with the naked eye. We are attracted by color, shape, texture, luster, and overall beauty of a specimen. En route to our collecting destination, we pay little attention to the hundreds of miles of exposed earth and country rock that passes by because it appears to be ordinary, uninteresting, or just plain ugly. Rubble.

Unfortunately, we are visually handicapped in our search for nature's gems. We cannot see the color in most of them. Our eyes are incapable of seeing all but a narrow band, wavelengths between 400 – 700 nanometers (nm), of the electromagnetic spectrum. This band contains all the colors of the rainbow and is referred to as the visible light spectrum. Invisible to us are the ultraviolet (UV) and infrared (heat) spectra. The ultraviolet spectrum contains wavelengths below 400 nm and is divided into four wavelength categories: longwave (400 – 350 nm), midwave (350 – 300 nm), shortwave (300 – 250 nm), and far ultraviolet (250 – 200 nm).

Fortunately, our visual handicap can be overcome through the use of shortwave and longwave UV lights that, as if by magic, turn rubble into gemstone. These lights were first used during World War II and the cold war era to locate tungsten (scheelite) for steel manufacturing and uranium ore minerals for nuclear weapons production. Both of these substances are highly fluorescent. During the second half of the twentieth century, high-tech prospectors armed with portable UV devices and Geiger counters, in search of new precious commodities, waited for nightfall to venture forth and light up the desert rubble and precious metal mine dumps abandoned by their nineteenth century pick and shovel predecessors.

New industries spawn new hobbies. Tungsten and uranium ores not only proved to be industrially useful, but they also fluoresce beautifully. Prospectors discovered that many other "useless" rocks and minerals were wonderfully entertaining when fluoresced. As UV lights became available for purchase by the general public, rock and mineral collectors found that fluorescents added a new and exciting dimension to their hobby. In fact, fluorescent collecting is a rapidly growing specialty hobby. Interest in fluorescence continues to grow among mineral collectors, not only for aesthetic reasons, but because fluorescence is an important mineral identification characteristic. And, modern mineral collectors are no longer merely "rockhounds", but fairly knowledgeable amateur mineralogists. One does not have to understand the scientific basis of fluorescence to enjoy the phenomenon. But, a basic knowledge of what happens when you pass your UV device over a piece of rubble and it suddenly glows like a blast furnace certainly enhances the experience.

Every mineral has its own chemical, atomic, and crystal structure. Fluorescence in a mineral depends upon whether or not its structure contains a fluorescent activator. Ideally, a mineral's chemical composition is pure and its crystalline structure is perfect. Thankfully, in the real world, chemical composition is often contaminated with foreign substances and crystallization is deformed by environmental influences. Although, some minerals are self-actuating like scheelite and uranium, most require a defect or impurity to activate fluorescence. Without venturing into graduate level atomic physics, suffice it to say that fluorescence occurs when electromagnetic energy from your UV light strikes a rock and is absorbed and altered by an activator. Some of the energy will be radiated back in the form of heat (infrared) and the rest as light. Because some of the energy has been converted to heat, the wavelength of the remaining light will be longer. If its wavelength is between 400 – 700 nm—voila! Fluorescence. Many more minerals fluoresce under shortwave than midwave or longwave UV light. Shortwave usually reveals bolder, brighter, and more contrasting colors. Perfectly structured and uncontaminated minerals will reflect the light from your UV device unaltered resulting in no fluorescence.

Fluorescence is caused by three types of activators. These are defect, intrinsic, and impurity. Defect activators are so-named because fluorescence is caused by a faulty atomic structure. When atoms are misplaced, misaligned, or are missing altogether, fluorescence may result. A few minerals, including even synthetic specimens, are consistently fluorescent regardless of where they are found. These are the intrinsically activated minerals. Scheelite is activated by its blue tungsten ion $(WO_4)^{2-}$. The molybdate ion $(MoO_4)^{2-}$ activates powellite a pale yellow color. The uranium group of minerals fluoresces yellow-green because of its uranyl ion $(UO_2)^{2+}$. Borate minerals, of which there are over 50, are intrinsically activated white, yellow, and orange.

Impurity activators account for the largest number of fluorescent minerals. As if gemstones such as emerald, spinel, ruby, garnet, and topaz are not beautiful enough, the inclusion of less that 1% chromium will cause then to glow a deep fiery red under long wave UV light. The disulfide ion S_2 causes yellow to red-orange fluorescence in members of the sodalite, scapolite, and cancrinite mineral groups. Rare earth elements used as activators in color television screens and other commercial applications also activate several minerals such as barite, apatite, anhydrite, fluorite, and zircon. The red fluorescence in calcite, yellow in fluorapatite, yellow-orange in halite and the famous multi-colored zinc ores of Franklin, New Jersey are activated by divalent manganese (Mn^{2+}). The uranyl ion that is the intrinsic activator in the uranium group also acts as an impurity activator causing fluorescence in hyalite opal, adamite, and chalcedony. Although activated by a uranium ion, these minerals pose no health hazard and are safe to collect.

However, the uranium minerals, of which there are over 170, require special handling because they are radioactive, toxic when ingested, and produce radon gas. There are two categories of uranium minerals: tetravalent and hexavalent. The tetravalent group contains uraninite and coffinite, the chief uranium ores found in the sedimentary rocks of Arizona's Colorado Plateau. The tetravalent minerals do not fluoresce. Most of the hexavalent minerals, those containing the uranyl ion, fluoresce brightly. Hexavalent minerals are the secondary uranium minerals that form in the oxidation zone that overlays uranium ore bodies. Using UV lights, uranium prospectors can detect the presence of underlying uranium ore bodies by detecting the fluorescent secondary uranium minerals on the ground. If you collect these dangerous minerals, consider storing them in lead lined containers or in a safe place out of doors out of the reach of pets and children. Better yet, although inconvenient, enjoy viewing them in their natural location and leave them where you find them.

Fluorescent minerals are best viewed in the dark. Even though fluorescents are constantly radiating UV light, the presence of even a little visible light energy can overpower the standard 12 volt UV lamp. Indoors, fluorescent viewing is easily done by waiting for nightfall or using a windowless room. Collectors who have chosen to come out of the closet have rigged up portable darkrooms using ponchos, boxes, blankets, and other contraptions for daytime fluorescent prospecting. These efforts can be moderately successful. Outdoor collecting is best done on cloudy moonless nights since moonlight and even starlight can interfere with fluorescent viewing. However, stumbling around among the cacti and pricker bushes in rough unfamiliar terrain in the dark can be hazardous. Portable battery powered UV lamps neither cast light very far nor illuminate your surroundings very well. When placing your hands and feet in ill-lit places, remember that unfriendly creatures like to prowl at night. Some of them even fluoresce. Scorpions glow a brilliant white under your UV light. So, be careful what you rush to pick up. A safer prospecting strategy is to reconnoiter the area in daylight and collect likely samples to view at home. If your samples do fluoresce, then after having familiarized yourself with the lay of the land, you can plan a safer nighttime expedition.

The colors of the UV spectrum are the same as those of the visible light spectrum—red, orange, yellow, blue, green, and violet. These colors, or hues, are characterized by saturation and brightness. The darker the surroundings, the more pronounced these characteristics become. Color terms are more descriptive than scientific. So, consider a pure hue as a departure point or a standard to measure fluorescent colors against. As you shine your UV light on a dark night, you will see gradations of color and varying degrees of radiation intensity. Variances in the same color, red for example, are a function of saturation. One rock may fluoresce pure red, the standard hue. Another that is less saturated may appear to be "pinkish". Other descriptive names such as light-orange, yellowish, blue-white, etc. are frequently coined to describe less than fully saturated fluorescent colors. When describing fluorescent colors, the artist's palate is sometimes more useful than the visible color spectrum. Brightness is the perception of power or intensity. Some rocks, and scorpions, will shine brilliantly at a considerable distance from your UV light. Others will manage only a dim, feeble response even when very close to your light. Those whose work demands precise analysis and standardized nomenclature use color atlases such as *The Munsell Book of Color* and scientific color and brightness measuring equipment. Since most readers will probably not have access to such scientific tools and references, fluorescent colors in this book are described in familiar and analogous terms such as beet-red, canary yellow, sky-blue, etc.

Frequently when viewing a fluorescent mineral you will notice that the color is not uniform or consistent throughout the piece. Even though a specimen may consist of only a single mineral species, fluorescence may be confined to only a certain portion or zone of the specimen. Much of the course chalcedony in Arizona, for example, will fluoresce a brilliant emerald green around the edges and surfaces of the piece rather than from deep within it. Some minerals present a speckled or mottled fluorescent appearance. Some minerals that have good to perfect crystal cleavage such as calcite, fluorite, and selenite display fluorescent geometric zones or faces confined within the piece by internal crystal structures. Often, rocks in Arizona will fluoresce orange on the outside. This is because the rock has been coated with caliche, a common sediment in southern Arizona.

Some fluorescent aficionados recommend testing everything you can get your hands on because fluorescent minerals can be almost anywhere. However, in the interest of saving time, knowledge of the general characteristics and causes of fluorescence will make your prospecting efforts more productive. Minerals with a metallic luster are much less likely to fluoresce than those with greasy, resinous, vitreous, and adamantine lusters. There is a correlation between electrical conductivity and fluorescence. Minerals that do not conduct electricity are more likely to fluoresce than those that do. Metals rarely fluoresce. Therefore, elements such as copper, gold, silver, iron, nickel and minerals such as galena, magnetite, pyrite, chalcopyrite, and hematite will display little if any fluorescence.

TABLE 17
Common Fluorescent Minerals

Mineral	Fluorescent Color	Mineral	Fluorescent Color
Adamite	Green	Glauberite	Blue
Agate	Yellow-green, cream	Gypsum	Green, pale yellow
Albite	Orange	Halite	Red, green
Alexandrite	Red	Hanksite	Pale blue, green
Alunite	Blue, orange (lw)*	Hemimorphite	Green, pale orange
Amazonite	Green	Howlite	Orange
Amber	Yellow, light green	Hydrozincite	Blue, orange, white
Amethyst	Deep blue	Kunzite	Reddish-yellow
Andalusite	Pinkish	Kyanite	Greenish-white
Anglesite	Yellow-green	Leadhillite	Yellow-orange
Apatite	Orange, yellow-green	Lepidolite	Pale green
Apophyllite	Yellow	Marble	Pink
Aragonite	Yellow, green, white	Opal	Green, blue, white
Axinite	Red	Powellite	Yellow
Barite	Orange, yellow, blue	Scheelite	Blue, white, yellow
Beryl	Yellow, orange, white	Selenite	Yellow
Calcite	Yellow, green, blue, red, orange, cream	Serpentine	Yellowish, cream
		Spinel	Red, yellow-green
Caliche	orange	Spodumene	Orange
Celestite	Yellow-green	Tourmaline	Yellow, lavender (lw)*
Cerussite	Yellow, blue, orange	Uranophane	Weak yellow-green
Chalcedony	Green, yellow, blue, orange	Wavellite	Blue
		Willemite	White, green
Corundum	Orange, red (lw)*	Witherite	White, blue
Diamond	Blue, green, red	Zircon	Orange
Fluorite	Green, blue, yellow		

Longwave ultraviolet light.

The Fluorescent Collecting Sites

SITE 31

Fluorescent Calcite near Cottonwood Gulch

Difficulty scale: 4 – 3 – 5   Seasons: Fall, Winter, Spring
Global Positioning System Coordinates: 34° 01' 45.7" N, 112° 11' 02.2" W *
Geology: Pliocene-Middle Miocene Volcanic Basalt, Tuff, Conglomerate
U.S Geological Survey 7.5 Minute Topographical Map: Black Canyon City

FROM INTERSTATE 17 BETWEEN Black Canyon City and New River, go west on Table Mesa Road (exit 236). At the west end of the Table Mesa Road overpass above I-17, turn north at the Frontage Road sign. "Frontage road" is rather a misnomer as it parallels the freeway for only a short distance before turning north-west toward the Bradshaw Mountains. Follow this road 1.7 miles to a fork. Bear right (north-west) go 2.8 miles, ford the Agua Fria River, and turn right at the top of the little hill on the other side of the river. The collecting area

The road cut collecting site at the crest of the hill.

is a road cut at the crest of a small hill 2.1 miles down this road. Unfortunately, the roads in this area are neither named nor marked.

On both sides of the road at the site, large, crumbly, igneous rocks are exposed by the road cut. Large veins of yellowish calcite lace through the host rock often filling vugs with crystalline patterns. This calcite fluoresces a medium-bright white. It also phosphoresces for a few seconds after you remove your UV light. The large rocks can easily be broken up with a sledge or heavy hammer. Strike them gently with only enough force to crack them. Then, separate and carefully trim the pieces so as to preserve the fragile crystalline structures.

G.P.S. coordinates taken at site.

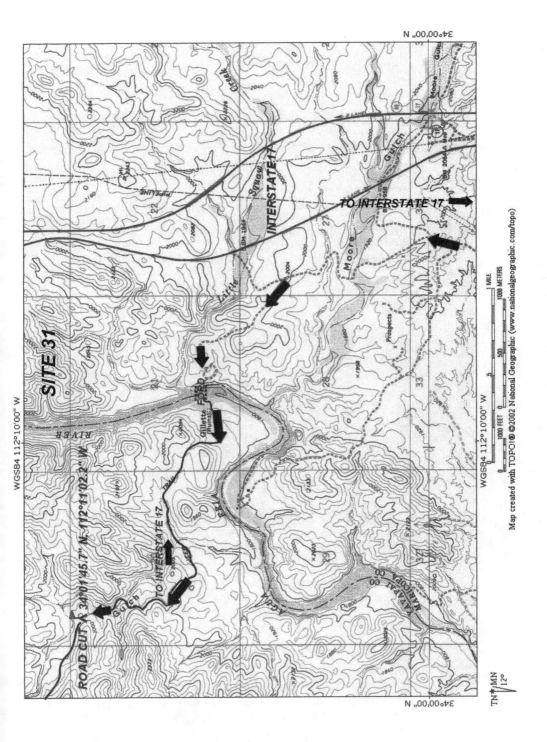

SITE 38

Fluorescent Geodes and Chalcedony at the Maggie Mine

Difficulty Scale: 5 – 5 – 5   Seasons: Fall, Winter, Spring
Global Positioning System Coordinates: Given With Each Location
Geology: Pliocene-Middle Miocene Igneous Basaltic Lava Flows
U.S. Geological Survey 7.5 Minute Topographical Map: Rawhide Wash

The road leading to the Maggie Mine sites showing the black cliffs and the entrance to Maggie Canyon.

FOLLOW THE DIRECTIONS ON PAGE 128 to the intersection of Signal and Alamo Crossing Roads. Turn left (south) on Alamo Crossing Road, drive 5.5 miles, and turn left (east). As you turn east toward the black cliffs ahead of you, you will see the entrance to Maggie Canyon (location no. 1) to the left (north), an open pit mine tailings slide directly ahead of you, and a second tailings pile (location no. 2) to your right (south). The topographic map, last updated in 1968, is somewhat confusing at this point. You will encounter several roads leading off in all directions that do not appear on the map. From the Alamo Crossing Road, drive .7 mile to a triangle intersection and bear right. Continue on another .9

mile to an intersection just below the first tailings slide. The road leading down the steep hill to the right goes southeast to the open pit excavations containing the quartz geodes (location no. 2). To reach location no.1 containing the fluorescent chalcedony, turn left (north) and drive .8 miles beyond the intersection to the entrance to Maggie Canyon.

Location #1, Maggie Canyon, G.P.S. Coordinates:
34° 20' 42.1" N, 113° 39' 14.2" W

The entrance to Maggie Canyon

As you turn the corner and enter Maggie Canyon you will soon come upon an immensely steep tailings slide on your right. This is site no.1. It is probably wise not to try to climb too far up this slide. You could cause a landslide. There is plenty of material to look through at the bottom part of the slide where it is less steep and more stable. Look for layers of botryoidal gray-blue chalcedony on very hard, heavy, dense, extrusive, igneous rock. This host rock has a fluid appearance, like taffy, displaying various black toned layers and swirls. This formation appears to have been a rapidly cooled molten flow. Some of the swirls are jet black and glassy like obsidian. Others are lighter and duller like rhyolite. White calcite crystals formed on top of the chalcedony. The chalcedony fluoresces a very pretty bright emerald green. Unfortunately, the calcite does not fluoresce.

Location #2 G.P.S. Coordinates:
34° 19' 55.3" N, 113° 38' 57.0" W

Location no. 2 is .49 mile from the intersection and is just below and to the south of the first tailings pile. You can see the excavations from the intersection.

Go down the steep hill and follow the road around in a southeasterly direction until you come upon the excavations on your left. This road is considerably rougher than any of the others in the area. The little geodes here look like small eggs. They have very thin white crystally shells that will break easily. Some are oblong and most have one side that is flatter and thicker than the other. All are hollow and some are so light that they will actually float in water. You will discover some that have weathered out and are loose in the soil, but most of them are contained in a soft, crumbly, salmon colored, muddy-looking igneous rock. This formation was probably near the top of a flow where gas bubbles penetrated the rock leaving it very porous like a sponge. The largest bubbles left room for the geodes to form. Look in the sides and bankings of the quarries and trenches as well as the rocks that have been dug out of them. The rock formation here is studded with geodes. The formations inside the geodes vary widely. Sharp, bright, clear quartz crystal needles; clean, white, drusey interiors; undulating chalcedony flows; and snow-white calcite crystal clusters are typical of the types of formations present in these geodes. Most will fluoresce a bright green.

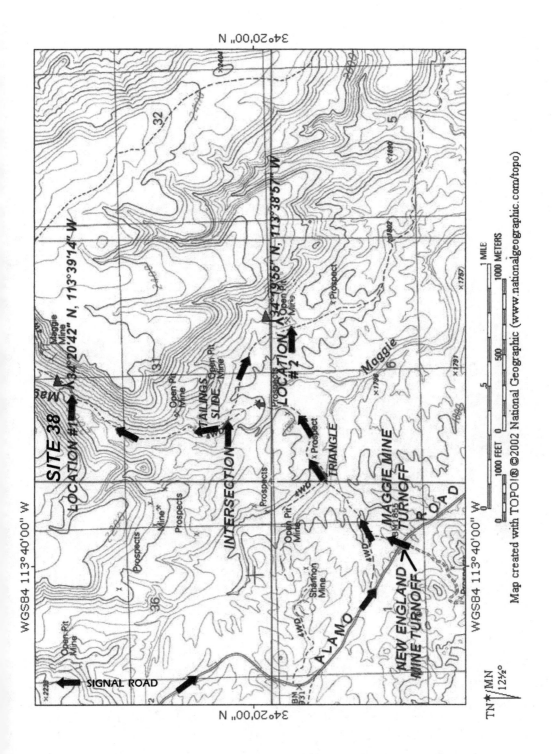

SITE 40

Fluorescent Calcite at the Scott and Black Pearl Mines

Difficulty Scale: 3 – 1 – 1   Seasons: Fall, Winter, Spring
Global Positioning System Coordinates: Given with each location
Geology: Middle Miocene Lava Flows and Tuff
U.S. Geological Survey 7.5 Minute Topographical Map: Belmont Mountain

The Scott Mine.

GO TO THE INTERSECTION OF Vulture, Wickenburg, and Aguila Roads north of Tonopah. If you are coming from the south, leave Interstate 10 at exit 103 and go north on 339th Avenue 2.9 miles to Indian School Road. Turn left (west) on Indian School, drive 2 miles, turn right (north) onto 355th Avenue (Wickenburg Road), and travel 17 miles to the Vulture Mine-Aguila Roads intersection. If you are coming from Wickenburg, take U.S. Route 60 west out of town 2.3 miles and turn south-west onto Vulture Mine Road. Follow Vulture Mine Road 19 miles to the intersection. If you are coming from Aguila, follow Aguila Road south-east

26.3 miles to the intersection. When you reach the intersection, turn onto the maintained gravel road that leads south-west into the Belmont Mountains. Follow this road 1.7 miles and turn right (west) onto an un-maintained dirt road. This turnoff is just a few hundred feet beyond the entrance to a gravel pit. Except for a few semi-rough spots, the road out to the collecting areas could be traveled in a passenger car. The distance to the Scott Mine is 5.4 miles and the Black Pearl Mine is another .6 mile beyond the Scott Mine.

Approaching the entrance to the Black Pearl Mine.

Both mines are hard to see from a distance because they are vertical shafts sunk in flat ground capped with rotten wooden covers. When approaching the Scott Mine (G.P.S. coordinates 33° 40' 53" N, 112° 58' 40" W) look for a faint track on your right (north) at the 5.4 mile point. About 100 yards in from the road, you will see a few small concrete foundations and a few pieces of sheet metal on the ground. Palo Verde trees are growing around the mine shaft and in the tailings pile. The Black Pearl Mine, .6 mile farther down the road (G.P.S. coordinates 33° 40' 50" N, 112° 59' 28" W) is easier to spot. Look for two concrete gate posts and the collapsed remains of a small wooden building in front of the mine shaft. The name "Black Pearl" is inscribed on one of the gate posts.

The tailings piles at both locations are small, low, and thinly scattered over the ground. Look for 2 – 3 inch pieces displaying flat calcite surfaces. This material fluoresces a medium bright cherry-red color. Zones of fluorescence appear intermittently across the rock surfaces. Some rocks are partially encrusted with an orange-yellow fluorescing caliche. The combination of these colors is quite pleasing.

If there is a rainy winter, plan your trip to this area in late February–early March to take advantage of the spectacular flora that blooms in the canyons and on the north slope of the Belmont Mountains. If you follow the road from the Wickenburg-Vulture Mine-Aguila Roads intersection past the turnoff to the collecting areas, you will come to the Tonopah-Belmont Mine at Belmont Mountain. This was a famous micro-mount collecting mine. But now, in order to save us from ourselves, the tunnel entrances have been barricaded like Fort Knox. Nevertheless, the golden poppies, bright blue lupines, and other colorful wild blossoms, and the lush green grasses that flourish among the saguaros and other cacti provides an awesome floral display.

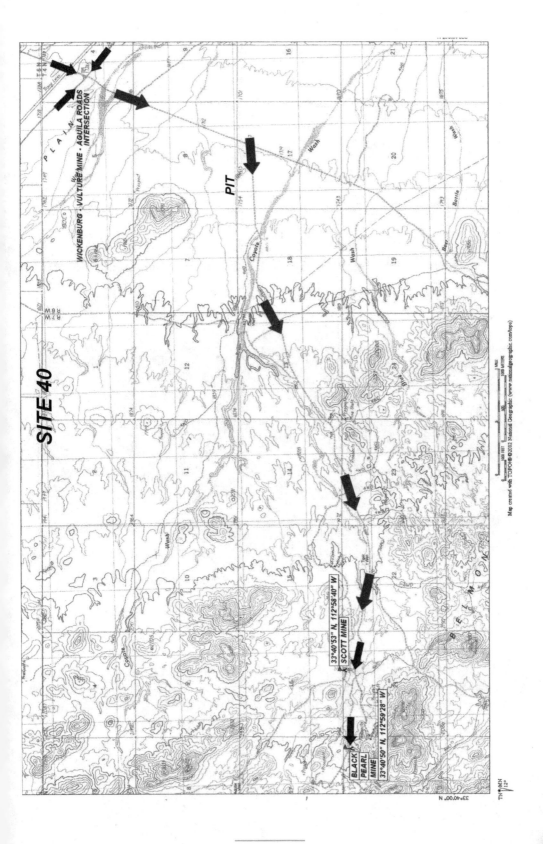

SITE 41

Fluorescent Calcite at the Black Silver Mine

Difficulty Scale: 4 – 3 – 3   Seasons: Fall, Winter, Spring
Global Positioning System Coordinates: 33° 17' 21" N, 113° 18' 11" W*
Geology: Middle-Late Pliocene Sedimentary Clay, Sand, and Gravel
U. S. Geological Survey 7.5 Minute Map: Columbus Peak

The turnoff to the Black Silver Mine from Charlie Russell Road.

FROM INTERSTATE 10, TAKE EXIT 81 and drive 11.9 miles south on Harquahala Valley Road to Baseline Road. Both of these roads are paved. Turn right (west) on Baseline, and drive 3.4 miles to 543rd Ave. Baseline becomes gravel just before you reach 543rd Ave. Turn left (south) onto 543rd Ave. The distance from here to the turnoff to the Black Silver Mine is 8.8 miles. After about 2 miles, 543rd Ave becomes Charlie Russell Rd. and curves around to the west at the 2.6 mile point. The turnoff to the mine is a faint track on your right (north). The distance to the small adit is .2 mile. Park here beside the trash heap and collect

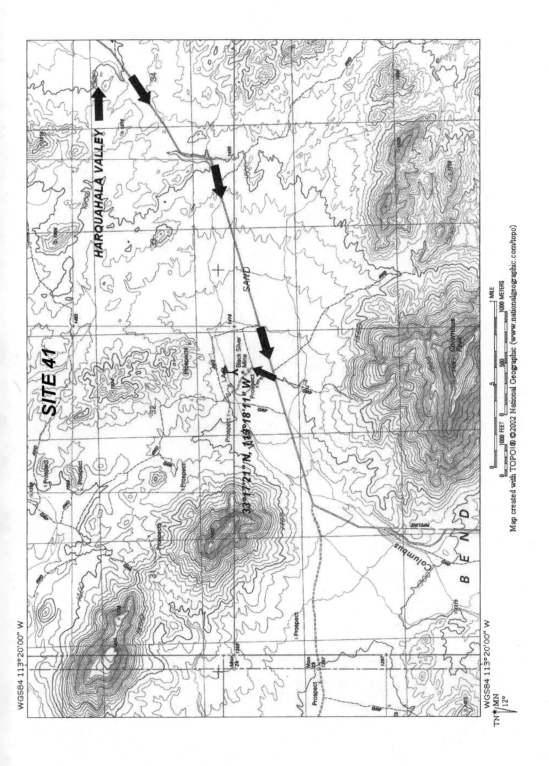

among the mine dumps on the right side of the road. Or, you can hike around the low hills and collect among the numerous little prospect holes that punctuate the landscape.

Look among the dark colored rocks for those that display a lot of calcite. The calcite is massive and dirty looking. There are no crystals or pleasing colors to recommend this calcite. There is nothing pretty about it until you shine your U.V. light on it. Then, it comes to life glowing a nice, rich red color.

*G.P.S. coordinates taken at the mine.

SITE 42

Fluorescents West of Morristown

Difficulty Scale: 5 – 5 – 4   Seasons: Fall, Winter, Spring

Global Positioning System Coordinates: Given With Each Location

Geology: Early Proterozoic Metamorphic Yavapai Supergroup and Pinal Schist

U.S. Geological Survey Topographical Map: Wickenburg SW and Vulture Mountain

FROM MORRISTOWN, which is roughly half way between Sun City and Wickenburg on U.S. Route 60, turn west on Gates Road. Go 2.3 miles to the east bank of the Hassayampa River. Look for the tire tracks in the sand of the dry river bed that end at a gate on the west bank. The distance across the river is .2 mile. Check the condition of the river bed carefully before attempting to cross. The Hassayampa is a major drainage system. Do not try to cross if water is present. Even days after it rains, sub-surface runoff can cause the river bed to be soft and muddy. In dry weather, the sand is usually compact enough to allow 2-wheel drive crossing. But, if enough traffic has passed to churn up the sand, then 4-wheel drive may be prudent.

The area between the Hassayampa River on the east and Vulture Mine Road on the west is primarily a gold mining district. Near Vulture Peak, several posted gold claims are still being actively worked. Fluorite, calcite, barite, lithium, silver, copper and a few other minerals were also mined here. There are miles of roads leading in all directions and lots of mines and prospects to explore. However, the area does not offer any lapidary material or mineral samples worthy of collecting. There are, however, several calcite dikes that were mined here. The tailings from some of these explorations contain some very colorful and interesting fluorescent minerals.

There are five mines in this district that yield particularly good fluorescent material. All five are calcite-fluorite dikes in igneous iron oxide rich matrix. The Big Spar and Mammoth Spar were primarily fluorite mines. The Queen of Sheba yielded gold, silver, and copper. Gold, silver, and manganese were extracted from the Newsboy Mine. At all these sites, look for rocks containing calcite and chunky pieces of solid calcite. Most pieces are dark, massive, and earthy looking. Some display semi-lustrous cleavage faces or have lighter colored calcite veins running through a dark rusty looking matrix. Occasionally, you may find a little white or transparent crystalline calcite attached to the darker, more massive calcite pieces. The calcite will fluoresce a very bright cherry red. Most pieces fluoresce multi-colored displaying blue and green fluorite, white scheelite, and orange caliche on a background of brilliant red calcite. Although ugly in daylight, this material is remarkably colorful under ultra-violet light.

Directions and distances to each of the collecting locations, except the Big Spar Mine, are measured from the gate at the west end of the Hassayampa River crossing. Be sure to leave the gate as you found it after you pass through. Although the roads to the collecting sites are used by cattle ranchers, they are rough and generally un-maintained requiring high-clearance vehicles. Many of the lesser used side roads are steeper and heavily eroded requiring 4-wheel drive. If you decide to explore these roads, be sure there is room to turn around if you suddenly encounter an impassible situation.

Location #1 G.P.S. Coordinates:
33° 51' 08.1" N, 112° 40' 07" W*

The "Main Road" leading west from the Hassayampa River showing Location #1 on the right.

The first collecting area, location #1, is visible from the gate. No name reference was found for this excavation. It is the hill top on the right (north) side of the road leading west up the slope from the gate. This road is labeled "the Main Road" on the map on page 351. Drive .2 mile up the Main Road until you come to a road on the right that leads up the hill. There is room to park and turn around if you choose to drive up the steep road to the top. The top and north-east side of the hill have been excavated and scraped off exposing black calcitic rocks. Look for those that have calcite layers sandwiched between thin limestone encrustations. Looking down into the ravine from the north side of the hill, you will see several rusty colored excavations. This is location #3.

*G.P.S. coordinates taken at collecting area.

Location #2, Newsboy Mine, G.P.S. Coordinates: 33°50'53.9" N, 112°40' 10.1" W**

View of the Newsboy Mine, location #2, from the Wash.

The entrance to location #2 is .15 mile west of the turnoff to location #1 on The Main Road, or .35 mile from the gate. Turn left (south), pass between the rusty iron gate posts, and immediately turn left (east) following the road around the side of the hill in front of you rather than taking the road leading up to the top of the hill. When you have gone around the hill, the road will then lead down into a wash. Turn right into the wash. The collecting site is about 200 yards up the wash on your right. The total distance from The Main Road is .5 mile. There is plenty of calcitic material in the wash to collect. Unless you are feeling adventurous, there is no need to scramble up the steep slopes above the wash.

**G.P.S. coordinates taken at collecting area in the wash

Location #3: the ravine at the bottom of the north side of location #1

The steep road leading to location #3.

No name reference was found for the location #3 excavations. The entrance to Site #3 is .3 mile west of the entrance to Site #2 on The Main Road or .65 mile from the gate. Turn right (northeast). The road angles and drops sharply to the right into the ravine. The distance to the digs is about 200 yards. The tailing piles here are dirty and colored a rusty red. The calcite here is typical of this district-dark and chunky, but you may also find a little clear angel-wing druse.

Location #4, Queen of Sheba Mine, G.P.S. Coordinates: 33° 52' 13" N, 112° 42' 18"W***

The turnoff to the right to location #4.

From the turnoff to Site #3, go 1.3 miles, or 1.95 miles from the gate, west on The Main Road to a fork. Bear right at the fork (north-west), drive 1.3 miles, and turn right (east) immediately after crossing a small wash. Go about 100 feet and turn left (north) up the hill. The collecting area is about 100 yards up the hill where you can park and turn around in a flat area by the digs. Do not try to

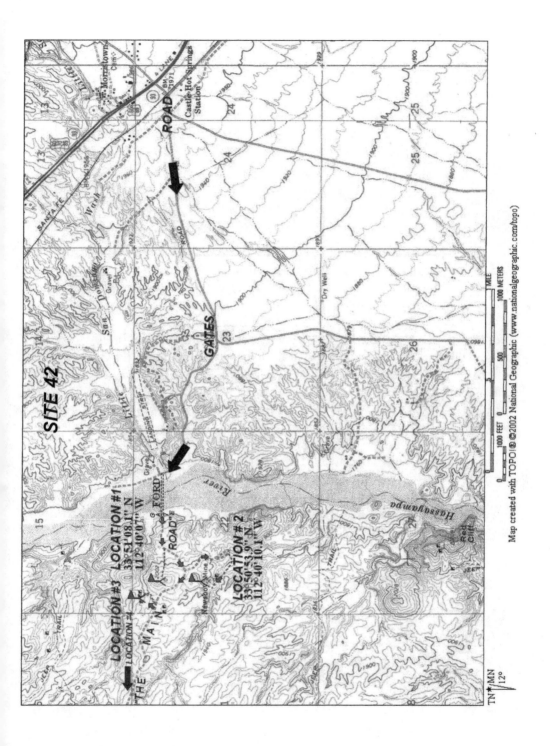

negotiate the steep rough road that goes farther up the hill from here. The material here is the most interesting and varied of all the five sites in this district. The rocks here contain some colorful chrysocolla and malachite as well as some small, shiny, metallic hematite crystal druse. The material fluoresces multi-color like that at the other four sites except it also displays an intense brilliant green between the hematite druse zones. This material fluoresces so brilliantly, that it is even visible in a lighted room. In the dark, it is spectacular.

***G.P.S. coordinates taken at the turnoff to the collecting area.

Location #5, Big Spar Mine, G.P.S. Coordinates: 33°53' 45.2" N, 112° 46' 42" W ****

Another .5 mile or so up the road from The Queen of Sheba is the Mammoth Spar Mine which is interesting to visit but yields little collectable material. It may be possible to continue on to Site #5, The Big Spar Mine, from here, but the going gets rough and 4-wheel drive is probably required beyond the Mammoth Spar. Instead, go back across the Hassayampa and go around through Wickenburg on U.S. Route 60 to the Vulture Mine Road. From the intersection of U.S. Route 60 and the Vulture Mine Road go 3.4 miles on the Vulture Mine Road and turn left (south) on the Vulture Peak Road. This is a well-maintained gravel road suitable for passenger cars. Go 2.25 miles on Vulture Peak Road and turn left (east), drive about 100 yards, turn left (north) and park by the big water tank. Collect the black calcite in the digs and tailings on the slope above. Be careful! There are some deep, unfenced, open, vertical shafts here. Like the Hotel California, you can check in, but you can't check out. Keep your eye out for clear calcite crystal clusters. They fluoresce a soft, light, baby blue on top of the bright red fluorescing dark calcite matrix. Immediately south of the Big Spar are several active gold mining claims. Best not to enter them without permission.

Approaching the turnoff to location #5 on the Vulture Peak Road.

****G.P.S. coordinates taken at collecting area.

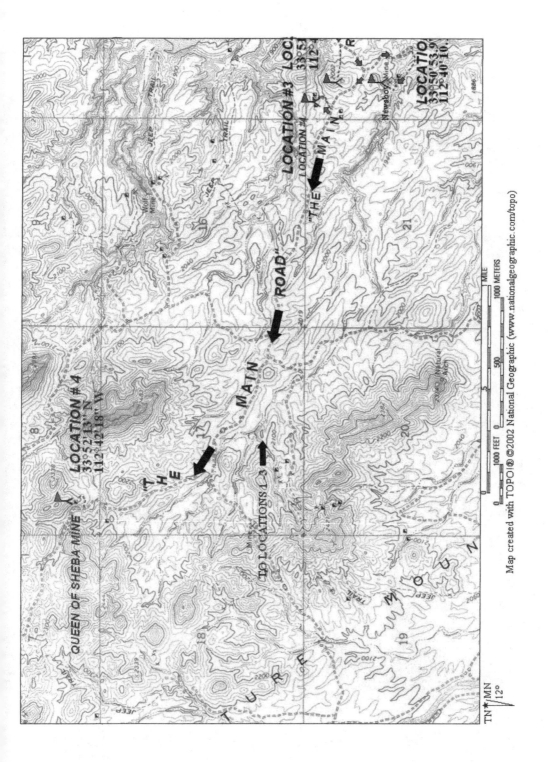

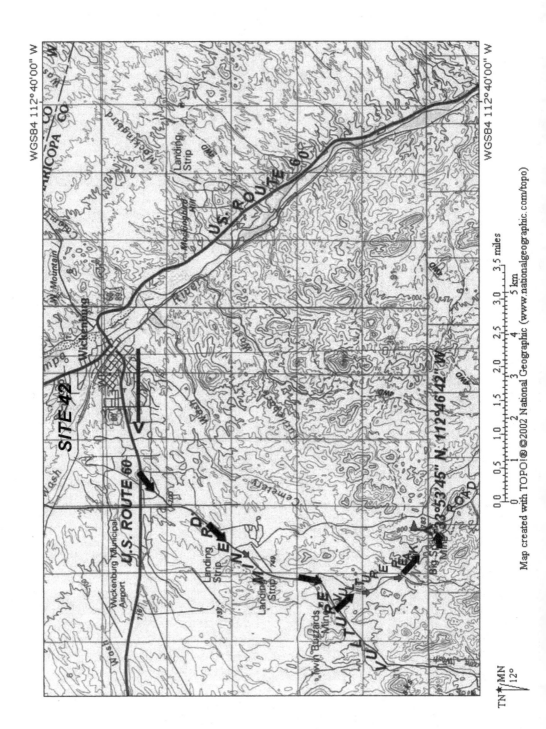

SITE 43

Fluorescent Calcite near Twin Buttes

Difficulty Scale: 5 – 3 – 1   Seasons: Fall, Winter, Spring

Global Positioning System Coordinates: 33° 47' 20" N, 112° 19' 28.7" W*

Middle Miocene-Oligocene Volcanic Lava Flows and Tuffs

U.S. Geological Survey Topographical Map: Baldy Mountain

FROM THE INTERSECTION OF I-17 AND STATE ROUTE 74, Carefree Highway, go east on S.R. 74 13.6 miles and turn left (south) on a rough un-maintained dirt road. This road is marked by a stop sign and a cattle guard and is between mileposts 17 and 18. If you are coming from the West, go 17.15 miles on S.R. 74 from the intersection of S.R. 74 and U.S. 60 and turn right (south). Proceed south 4.5 miles where you will come to a T intersection at the base of the north-east side of Twin Buttes. The collecting area is directly in front of you about 500 feet up the hill. On your left, will be a large flat graded area. If you follow the road on your right, a few hundred yards around the corner you will come to two holding ponds on your left. Instead, turn left at the T and go about 100 yards, turn right and almost immediately turn right again onto the road that leads you up to the collecting area. Look for a tall, skinny saguaro beside a pile of broken concrete.

The calcite here is black and white and lies scattered around the graded area surrounding the lone saguaro. The material is massive but displays well-defined crystal faces and chevron shaped cleavage plains. Pieces about six inches

The skinny saquaro beside the pile of broken concrete.

across are plentiful. The fluorescent surfaces of this material present a medium bright red appearance.

If you desire larger pieces, go left (south) at the T intersection about .1 mile where you will find several large calcite boulders on the left side of the road. If you feel energetic, you can break off larger pieces from these boulders with a sledge. More calcite is available farther down the road at the White Peak Mine which is actually not at White Peak, but just north of it and directly west of Twin buttes. To get there, continue southward down the road from the T intersection .6 mile and turn right (west) immediately before the bridge that crosses the canal. Follow this road 2.3 miles into the lower level of the White Peak Mine pit (33° 47' 01" N, 112° 21' .09" W). Look for chunky and crystalline calcite pieces in the bankings and tailing piles in and around the small open pit. Unfortunately, the material here fluoresces only weakly.

Much better material is available at the Prince Mine which is west of the White Peak Mine. You can see the road leading to it going up and over the ridge to the West. However, this is a rather difficult 4-wheel drive adventure. See page 358 for an easier route to the Prince Mine collecting site.

G.P.S. coordinates taken at the collecting area.

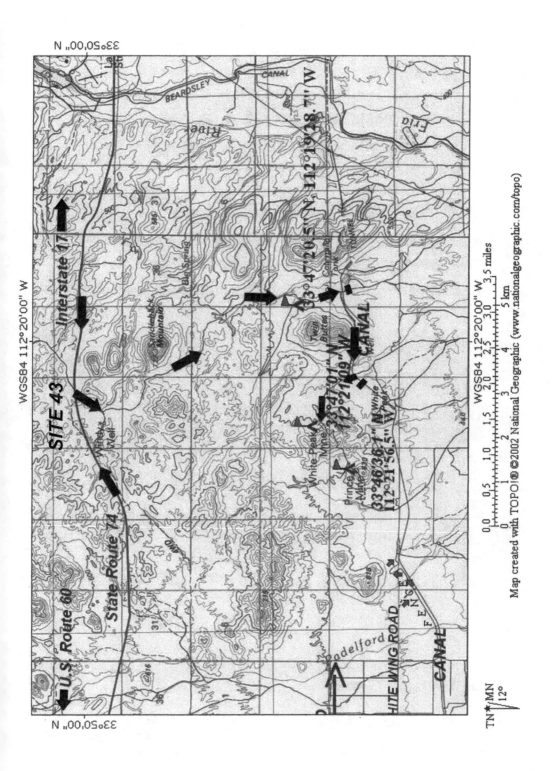

SITE 44

Fluorescent Calcite at the Prince Mine

Difficulty Scale: 4 – 5 – 3   Seasons: Fall, Winter, Spring

Global Positioning System Coordinates: 33° 46' 36.1" N, 112° 21' 56.5" W*

Geology: Middle Miocene-Oligocene Igneous Lava Flows and Tuffs

U.S. Geological Survey Topographical Map: Baldy Mtn

The road leading up Prince Peak to the Prince Mine.

IF YOU ARE COMING FROM PHOENIX, take Grand Avenue (U.S. Route 60) to 163rd Avenue and turn right (north-east). Go 5.7 miles, cross the canal, and turn right (east) on White Wing Road. If you are coming from Wickenburg on U.S. Route 60, turn left (east) on Center St. in Wittmann. Go .4 mile and turn right onto Dove Valley Road. You will still be going in an easterly direction. Drive 6.55 miles to 163rd Avenue. Dove valley Road varies considerably from paved to well-maintained gravel, to narrow un-maintained dirt, to well-maintained again. Turn right (south) on 163rd Avenue, go about 1 mile, and turn left (east) on White Wing Road. Follow White Wing Road 1 mile and turn right (south). Go .2

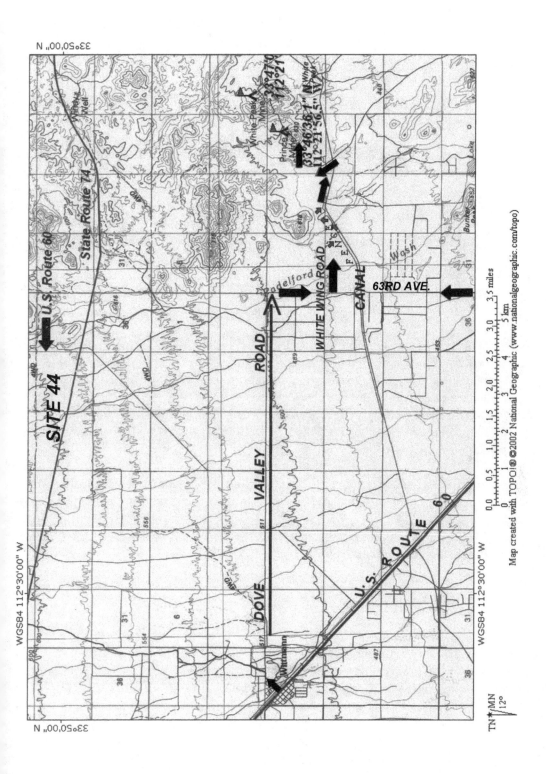

mile until you reach an opening in the fence. At this point, you will be behind a flood control dam the parallels the Central Arizona Project Canal. Check to be sure the road is not muddy or under water before proceeding. If the road is passable, turn left (east) and follow it as it snakes around the flood control dam system and off through the chollas and creosote bushes toward White Peak and Prince Peak. The road becomes so narrow that it begins to look like an ATV trail. But, if you can stand a little Arizona pin-striping on you vehicle, and one or two steep wash crossings, you can negotiate it successfully. After 1.4 miles, you will come to a cross road. Turn left (north) on this road and drive 1 more mile to the Prince Mine. This road will curve to the east as it makes its way up the north side of Prince Peak to the mine. At the 1 mile point, you will find the mine tailings and excavations on the right hand side of the road. All you have to do is park and start collecting. At the time of this writing, the area around White Wing Road was being readied for a housing development. It is unclear whether or not this route to the Prince Mine will still be open. If not, then go to the White Peak Mine (Site 43, page 355) and hike westward over the ridge and down to the Prince.

There is a huge amount of material here. Large piles of black and chocolate brown chunky calcite of all sizes line the uphill side of the road. Be careful as there are trenches and tunnels among the tailings. Most of the calcite is ugly, dark, and massive. There are a few pieces containing white zones and crystal faces that are rather nice. The real value of this material is its excellent fluorescent quality. When you turn your UV device on it, it fluoresces a very bright, intense, red color.

If you continue on the road through the Prince Mine to the east you can see where it curves up and over the ridge to the White Peak Mine. But unless you are a white-knuckle 4-wheel drive addict, follow the directions on page 355 to the White Peak Mine.

G.P.S. coordinates taken at the mine.

SITE 45

Fluorescent Calcite North of Lime Hill

Difficulty Scale: 4 – 3 – 3   Seasons: Fall, Winter
Global Positioning System Coordinates :32° 14' 18" N, 112° 55' 35" W*
Geology: Devonian Limestone
U.S. Geological Survey 7.5 Minute Topographical Map: Bates Well

View of snow white Lime Hill from the intersection of Bates Well Road and the turnoff.

FROM THE PLAZA IN DOWNTOWN AJO, follow State Route 85 2.6 miles south past the huge mine dumps to Darby Well Road. Turn right (west) on Darby Well Rd., drive 1.9 miles to Bates Well Road, and turn left (south). There is no road sign for Bates Well Road, but you can easily recognize it because it is a well-maintained road suitable for passenger cars. You will also see a sign announcing that this area is closed between April 30 and July 15 for antelope fawning. Follow Bates Well Road 7.3 miles and turn right (west) on the un-maintained dirt road that leads to Lime Hill. You will begin to see snow white Lime Hill rising up out of the desert all by itself about half way down Bates Well Road. You will come to a fork in the road after driving 1.1 miles. Bear right and continue for another 1.3 miles to a second fork. The road to the left goes up Lime Hill. Instead, Bear right and follow the road north .4 mile to the turnoff to the Growler Mica mine. The road will only take you about .1 mile before it becomes impassable at an overgrown wash. At this point, you will have to park and pick your way through the heavy underbrush in the wash. Once you reach the other side, the going gets much easier. All you have to do is walk across the open, level ground beside the low quarry dumps and pick out whatever you like.

The turnoff to the Growler Mica Mine.

The calcite here is much more interesting than the stuff at Lime Hill. It has dark and light colored patterns and displays fairly large crystal faces imbedded in massive calcite. It also forms clusters of parallel tubular structures like a bundle of straws. Unlike the Lime Hill calcite, some of this material will fluoresce, but not the typical red color. Instead, the tubular type fluoresces green and the rind on the black and white patterned type fluoresces orange. Randomly, zones of green fluorescence resembling colonies of green pin points will show up on some pieces. This may be willemite.

If you wish to explore the Lime Hill excavations, 4-wheel drive will take you half way up the hill. Otherwise, you will have to park and hike up. Trying to climb around on the tailings is unnecessary and dangerous. The tailings slides are very steep, unstable, and contain huge bone crushing boulders. The material here is pretty ordinary massive calcite adorned with only a few small crystals. Unfortunately, it does not fluoresce.

While you are in Ajo, it is worth your time to visit the little Ajo Museum in the old Indian mission church. From here, you can gaze down into the famous

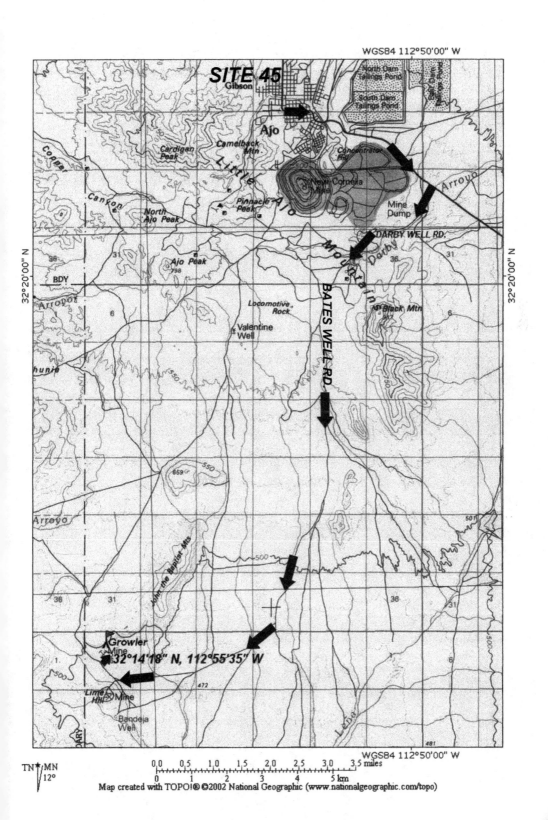

New Cornelia Mine open pit which is the type location for the mineral ajoite. In addition to the usual interesting local artifacts, the museum contains a respectable mineral collection assembled by the area rock club. Of special interest is the excellent display of New Cornelia Mine minerals. The ajoite, papagoite, and shattuckite specimens are excellent. The museum also houses a collection of documents, newspaper clippings, photographs, reports, etc. concerning local history, people, events, and mining activities. Places such as this provide an important source for original research which enriches our mineral collecting hobby.

G.P.S. coordinates taken at the Growler Mica Mine collecting site.

SITE 63

Fluorescents South of Cottonwood Canyon Road

Difficulty Scale: 5 – 5 – 3   Seasons: Fall, Winter, Spring
Global Positioning System Coordinates: 33° 11' 22" N, 111° 14' 50" W*
Geology: Proterozoic Metamorphic Yavapai Supergroup and Pinal Schist
U.S. Geological Survey 7.5 Minute Topographical Map: Mineral Mountain

THIS IS THE FIRST OF FOUR SITES in the aptly named Mineral Mountains. This mountain range is composed of early Proterozoic metamorphic rock-sedimentary and volcanic rock metamorphosed to gneiss and schist. The primary formations here are the Pinal Schist and the Yavapai Supergroup. Before proceeding to your destination, carefully read the notices posted in the parking area on the south side of the road because, you will be traveling through an Arizona Army National Guard live fire training area.

View of the road leading up to the collecting area from Cottonwood Canyon Road.

To reach this site from Florence Junction, (the intersection of State Route 79 and U.S. 60) drive south on S.R. 79 5.4 miles and turn left (east) on Cottonwood Canyon Road. If you are traveling from Florence, go north on S.R. 79 about 10.5 miles and turn right (east) on Cottonwood Canyon Road. Follow Cottonwood Canyon Road 5.9 miles and turn right (south) onto a rough un-maintained road that crosses the wash. Follow this road up and around a hill .5 mile to a place that you can turn around and park. As you walk farther down the road toward a small tailings slide up on the hillside ahead of you on the left, another small excavation will become visible on the hillside beside the road to your right. These appear to have been unsuccessful copper prospects. You will have to scramble up the hillside to reach the excavation on your left. There is a narrow little road on the right leading to the other excavation.

Some of the metamorphic rock here will fluoresce multi-color. Look in the mine rubble for rocks containing calcite which is visible and easily identifiable in the earthy looking metamorphic rock. Other fluorescent minerals tend to be present in the calcitic rock but are hard to detect with the naked eye. These other minerals fluoresce blue-white, green, and orange along with the red calcite. Given the history of the Mineral Mountain and Mineral Hill mining districts, the blue-white could be hydrozincite or scheelite, the green could be chalcedony or willemite. But, without further analysis, we can not be certain what these minerals really are.

* G.P.S. coordinates taken at the turnaround.

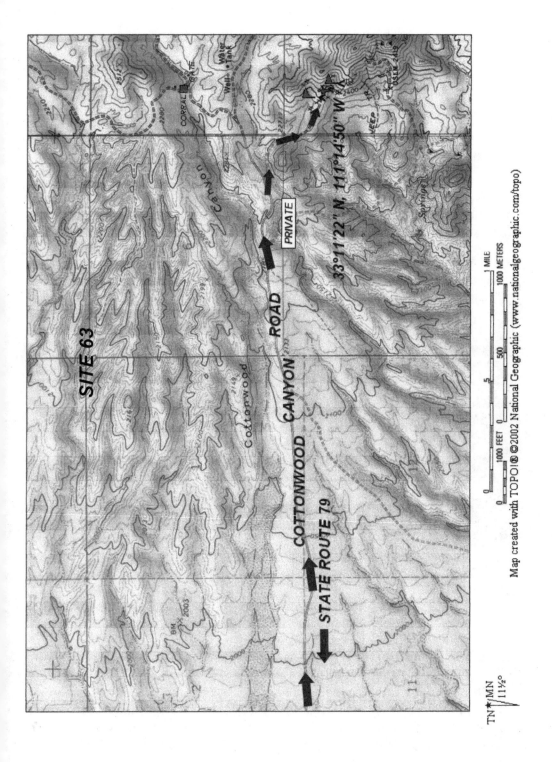

SITE 65

Fluorescents at the Ajax Mine

Difficulty Scale: 6 – 5 – 6   Seasons: Fall, Winter, Spring

Global Positioning System Coordinates: 33° 12' 27" N, 111° 11' 02" W*

Geology: Early Proterozoic Metamorphosed Yavapai Supergroup and Pinal Schist

U.S. Geological Survey 7.5 Minute Topographical Map: Mineral Mountain

View of the Ajax Mine from the road

THE AJAX MINE is just .5 mile down the road from the turnoff to the Woodpecker Mine, Site 64. For directions, see pages 180 and 181.

In the tailings that have fallen into the wash, search for rocks containing calcite. The mineralogy of this mine is quite complex and includes several different fluorescent minerals. U.V. light reveals reds, greens, blues, and oranges, of various hues and intensities. The best way to collect here is at night with a portable U.V. device. But, if you do it is probably a good idea to plan to camp overnight rather than attempting to negotiate the long, tedious, rugged road back in the dark. The other alternative is to gather a large variety of likely looking rocks for testing at home.

As a bonus, you can also collect amethyst here. On the top of the hill above the mine shaft and tailings pile, there is a tall, sharp reef of large, craggy, massive, metamorphosed volcanic rock similar to that found at the Woodpecker Mine. This rock is heavily veined with stripes and swirls of semi-clear quartz and hematite veins. Although, most of the rock here is very large requiring the use of a heavy sledge hammer, there is a tailings pile of smaller rocks suitable for cutting and polishing available in the middle of the reef. The veining in the interior of some of these rocks is amethyst. The combination of quartz, hematite, and amethyst colors and patterns make this rock an excellent lapidary medium.

G.P.S. coordinates taken at the mine.

SITE 66

Fluorescent Minerals South of Mineral Mountain

Difficulty Scale: 4 – 4 – 3   Seasons: Fall, Winter, Spring

Global Positioning System Coordinates: Given With Each Location

Geology: Early Tertiary-Late Cretaceous Igneous Granitic diorite and porphyry

U.S. Geological Survey 7.5 Topographical Map: Mineral Mountain

FROM THE INTERSECTION OF Price Road and State Road 79 in Florence, go east 7.9 miles on Price Road and turn left on the dirt road leading north. Drive .75 mile to a fork in the road. Bear right (north-east) and proceed 3.3 miles to a second fork. The road to the left (north) goes to the Oklahoma Mine and locations 4 and 5. The road to the right (east) goes to locations 1, 2, and 3.

The hills and ridges within an area 1 – 2 miles south and east of the Oklahoma Mine are heavily prospected. The rock here is early Tertiary-late Cretaceous granite, diorite, and near surface porphyry containing copper deposits. Past mining ventures have left holes and shallow excavations in almost every rock outcrop. Consequently, the area is fertile ground for rockhound exploration. The fluorescents found here are excellent and the quartz and amethyst crystal

The short, eroded road leading to the base of the knob at location #1.

veinings are well-formed and gemmy. This is a great area for daylight cross-country rock hunting on foot. And, if you have a portable U.V. light, it is a superior area for nighttime fluorescent prospecting.

Location # 1 G.P.S. Coordinates:
33° 09' 11.9" N, 111° 14' 0.01" W*

To reach location 1, bear right at the second fork and proceed .4 mile. The collecting area is the small volcanic knob across the wash on your left. The surrounding area is typical of an abandoned mine site; trash, ruins, prospect excavations, and tailings piles. You can drive down the short, eroded road and park in the wash at the base of the knob. If you scramble up the side of the knob, you will find the going steep and slippery. Beware of the half filled-in shafts and tunnels amongst the rubble as you ascend the slope. Actually, you do not have to climb very far up the slope. There is plenty of material to collect near the bottom. Look for pieces of rock containing black calcite. If you walk up the short road to the top, you will find large calcitic boulders. With a sledge, you can break off pieces of attractive, bright, shiny black calcite laced with bright, white, crystally quartz and calcite veins. The calcite at this location fluoresces a medium bright red in solid and pin point patterns.

Follow the road eastward a few hundred yards to a trench mine on you left (north). The rock here is darker than the surrounding area. Search the tailings in and around this trench for quartz crystal vugs contained within a light peppermint green and maroon streaked matrix. The vugs that are exposed are weathered and rust stained. By cracking the larger rocks open with a heavy hammer, you will expose fresh, bright, clean crystal colonies.

Location #2 G.P.S. Coordinates:
33° 09' 08" N, 111° 13' 28" W*

When you are finished here, continue up the road another .55 miles to Location 2. Beginning at the point where the narrow dirt road intersects from the left (north) up the hill in front of you for 100 yards or so, look for quartz crystal vugs in the volcanic matrix. Check the rock piles and ledges on both sides of the road and in the roadbed. As before, breaking the rocks will yield the best quality crystals. You might find a little amethyst here. Apparently, an amethyst vein runs beneath the road and a few pieces were plowed up when the road was graded.

Location #3 G.P.S. Coordinates:
33° 09' 12.5" N, 111° 13' 14." W*

Continue on another .3 mile down the road where you will encounter dumps of black and cloudy calcite rocks mixed with quartz. These rocks measure between 2 and 12 inches across. Collecting is simple. Just park on the road and pick what you want from the piles on both sides of the road. This rock fluoresces exceptionally well. It is reminiscent of the choice Hogan Mine material. It glows with brilliance and intensity. The primary colors are bright red and deep blue. In addition, some pieces show traces of white, orange, and green. A chunk of this material will provide a rich addition to any fluorescent collection. On the hillside about fifty feet above the right (east) side of the road, you will see a dirt pile. You can follow the road a few hundred yards up and around to the top of this pile. You will probably need four wheel drive for this steep, rutty, little stretch of road. Or, you can simply hike the short distance up the hill. Once there, you will find a mine shaft with a tailings pile of black calcite laced with vugs and veins of clear quartz crystals, amethyst crystals, white angel wing calcite, and a little white barite. Dig into the tailings and break open the bigger rocks to reveal the crystal chambers inside. This material fluoresces red, blue, green, and white.

Location #4 G.P.S. Coordinates:
33° 09' 27.2" N, 111° 14' 17.6" W**

To reach location 4, return to the second fork in the road and take the road that leads north toward the Oklahoma Mine. Go north .5 mile until you are opposite a small adit on your left at the base of a dark hill that looks like an inverted cone. To the left of the adit, there is a path leading up the hill to another tunnel on the south side of the hill. Bring a heavy hammer. Look for heavily quartz veined rocks littering the hillside. Break these to find the quartz and amethyst veins and vugs inside. The crystal veins are usually horizons where

Fork #3 overlooking the Oklahoma Mine area.

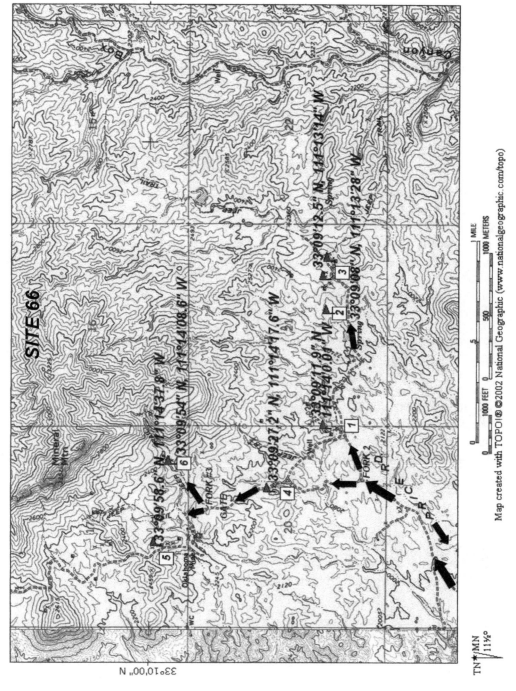

crystals have formed on both the top and bottom opposing each other like a set of dentures. Sometimes they have grown together forming a solid vein. Amethyst veins and crystals that have been exposed to the sun have faded and become colorless. By cracking rocks displaying clear crystals, you may discover pretty blue amethyst interiors. On the difficulty scale, this hillside is a steep and slippery 6.

Location #5 G.P.S Coordinates:
33° 09' 58.6" N, 111° 14' 37.8" W*

To get to locations 5 and 6, continue northward up the road from location 4, through a gate, .4 mile to fork number 3. At this point, you are at the crest of a hill overlooking the Oklahoma Mine area. You will see a number of roads meandering across the landscape in front of you. The left fork curves westward, goes past the Oklahoma Mine, and then curves northward around to location 5. The total distance is .7 mile. The right fork takes you to location 6. At location 5, look for excavations and tailings piles on the right (east) side of the road containing pieces of black calcite. Although similar in appearance to the calcite found at locations 1 and 3, this material fluoresces a lighter orange-red color. There is nothing much to collect at the Oklahoma Mine except for some specimens of thin malachite and chrysocolla crusts.

Location #6 G.P.S. Coordinates:
33° 09' 54" N, 111° 14' 08.6" W***

To reach location 6, return to fork number 3 and take the road leading east .4 mile and park where you can turn around. The collecting area is a small prospect hole a few hundred feet to the west on top of the little hill above you. There is a rock outcrop here that is laced with quartz crystal vugs. You can look through the rubble left from the prospect hole or, if you feel energetic, you can attempt to enlarge the hole with crow bar and sledge hammer in search of fresh material.

*G.P.S. coordinates taken at collecting locations.
**G.P.S. coordinates taken on road opposite adit.
***G.P.S. coordinates taken on road at turnaround.

SITE 71

Fluorescent Scheelite at the Tungsten King Mine

Difficulty Scale: 5 – 6 – 4   Seasons: Fall, Winter, Spring
Global Positioning System Coordinates: 32° 04' 22" N, 110° 09' 10" W*
Geology: Early Proterozoic Metamorphic Yavapai Supergroup & Pinal Schist
U.S. Geological Survey 7.5 Minute Topographical Map: San Pedro Ranch

THERE ARE TWO ROUTES TO THIS SITE. If you are traveling eastward on Interstate 10 from Tucson, take exit 306 near Benson and go north on Pomerene Road 3.1 miles to Pomerene. Turn right (north) in Pomerene on Cascabel Road, drive 5.9 miles, and turn right (east) at the top of the hill onto the road leading down into Tres Alamos Wash. You will have to drive about .1 mile before Tres Alamos Wash comes into view. Drive up the wash 3.5 miles and turn right (south-east). This road will take you up and out of the wash. Follow it 1.8 miles to a T at a corral. Turn left (east) and go 1 mile where you will come to a road on your right (south) that leads to the Z R Hereford Ranch. From this point continue eastward, past the turnoff to the ranch, another 1.4 miles to a fork. Bear right (east), drive 1.7 miles toward Clark Canyon and park. From here, you can see the road continuing up into Clark Canyon to the Tungsten King adit. Do not attempt to drive up this road! It is steep, narrow, and there is no room to turn around once you reach the top. Unfortunately, unless you have an ATV, the only way up is to hike.

If you are traveling westward from Wilcox on Interstate 10, go north on Sibyl Road, exit 312, 3.4 miles to the Z R Hereford Ranch corral. Drive slowly, watch out of ranch animals, and respect private ranch property. Sign in at the log box at the corral gate. Proceed through the corral, go west .6 mile, and turn right (north) on the road that leads up and out of the wash. Follow this road .8 mile up to a pipe gate. From the gate, go .6 mile to a fork. Turn left (north) at the fork and follow the road 1.9 miles to a large corral. At the left (west) end of the corral follow the road through a wire gate. Do not turn right and follow the road eastward through the pipe gate even though it looks like it leads in the right direction. After going westward through the wire gate, follow the road 1.3 miles north-west to a T even though it seems to be the wrong direction. On your left

The lower collecting area and the steep narrow road leading to the mine as seen from the parking area.

will be the road leading westward back to Tres Alamos Wash. Turn right (east) and go 1.2 miles to the fork that leads 1.7 miles to the parking area at the entrance to Clark Canyon.

En route to the Tungsten King, you will be traveling through state trust land. To do this legally, you are required to have a permit which you can acquire at the Arizona Public Lands Information Center, 222 N. Central Ave., Phoenix, AZ. 85027 (602-417-9300). The fee is $ 15.00 a year for an individual or $20.00 a year for a family.

There are two scheelite locations here. The first and most difficult to reach is the main mine at the end of the steep, narrow road that leads up to the upper reaches of Clark Canyon. You can see it rising steeply up the hillside as you approach the parking area.

The hike up the road to the top is long, steep, and arduous. Part of the way, the road is actually paved with rough cement. But, sand and small rocks litter the pavement making it very slippery. When you finally arrive at the top and catch your breath, you will find tailings on both sides of the canyon wash. If you have a portable U V light, bring it with you. All the rocks here look pretty much alike. It is much easier to select only the best specimens at the top, instead of highgrading and discarding half the rock you lug down at the bottom.

There is a second location in the lower level of Clark Canyon which is at about the same elevation as the parking area. To reach it, go to the small water tank just a few hundred feet up the road from the parking area. From here, hike eastward into the canyon to the edge of the canyon wash and follow it upstream

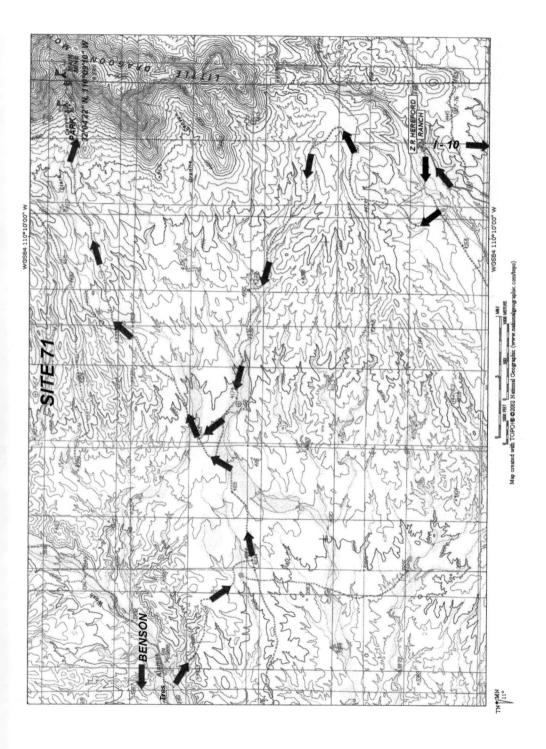

to the excavation and tailings on top of a small granite outcrop. The distance from the tank to the tailings is about .25 mile. Wear long pants and sleeves because you have to fight your way through some thick pricker bushes.

At the time of this writing, the authorities were considering closing this area to public access to protect an endangered species of bat that lives in the mine shaft. Apparently the bats are migratory and spend only part of the year at this location. Whether or not the area will be closed and if so for what part of the year remains uncertain. If the area is closed, hopefully it will only be during the period when the bats are present and the tailings at the bottom of the canyon will remain accessible.

*G.P.S. coordinates taken at the parking area.

SITE 78

Fluorescent Calcite at the Crook Tunnel

Difficulty Scale: 3 – 4 – 3   Seasons: Fall, Winter, Spring

Global Positioning System Coordinates: 31° 20' 16" N, 109° 49' 37" W*

Geology: Cretaceous-Late Jurassic Sandstone and Conglomerate

U.S. Geological Survey 7.5 Minute topographical Map: Bisbee SE

GO TO BISBEE JUNCTION SOUTH of the town of Bisbee and east of the town of Naco. Drive east on Border Road 4 miles and turn left (north) on a narrow, unmaintained, dirt road. Go .2 mile to a fork. Turn left (west) and follow the road parallel to the railroad track another .2 mile to the turnaround at the end of the road and park.

From the parking area, walk northward through the hole in the barbed wire fence and cross the railroad tracks to the other side where you can access the large rubble pile. As you cross the tracks, you will be able to see the Crook Tunnel on your right (east). Although, from the look of the rusty tracks and the rotten railroad ties, this railroad grade appears to be out of service, it is probably prudent to stay out of the tunnel. In the rubble piles, look for chunks of massive dark colored calcite and bronzy colored aragonite encrustations. Although there

View of the collecting area across the tracks from the turnaround.

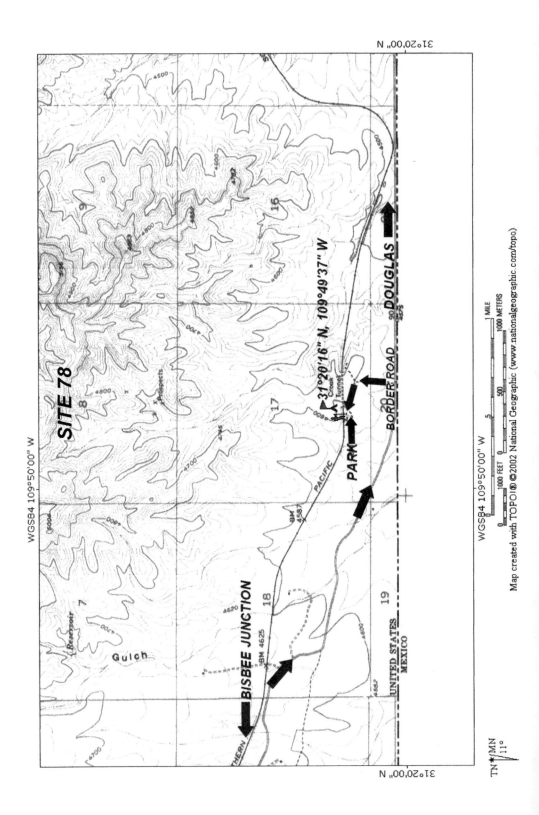

are plenty of 2 – 3 inch pieces to collect here, you may want to trim up some of the larger rocks with you rock pick. The calcite fluoresces a bright red and the aragonite fluoresces a mild white color.

As you travel Border Road eastward toward Douglas, there are several other roads that lead northward to the railroad grade that parallels Border Road for several miles between Crook Tunnel and State Route 80. You can collect more fluorescent calcite and aragonite at the railroad cuts along this stretch of track. Be vigilant for illegal immigration and Border Patrol activity in this area.

*G.P.S. coordinates taken at parking area.

SITE 80

Fluorescents along the San Francisco River Road

Difficulty Scale: 2 – 2 – 1   Seasons: All
G.P.S. Coordinates: 33° 06' 21" N, 109° 17' 29" W*
Geology: Paleozoic Sedimentary Rocks and Volcanic Rocks and Flows
U.S. Geological Survey 7.5 Minute Topographical Map: Clifton

From U.S. Route 191 in Clifton, turn right (east) at the American Legion building onto San Francisco River Road. Do not cross the bridge immediately in front of you. Instead, turn left (north) staying on the west side of the river and go .55 mile to the Polly Rosenbaum Bridge. Cross the San Francisco River here. Distances to the collecting points upstream are measured from the bridge.

The geology along the San Francisco River Valley in the area of Clifton-Morenci is varied and complex. The landscape is characterized by Paleozoic brown and gray sandstone and green shales overlain by light to medium gray limestones and aragonite from the Mississippian, Devonian, and Cambrian periods. Contained within these formations are Miocene and Oligocene igneous intrusions, lava flow overlays, and welded tuff. The strata have been significantly faulted, tilted, and contorted. Consequently, all of this geologic phenomena has created lots of mineral collecting interest and opportunity.

Looking north on the San Francisco River Road.

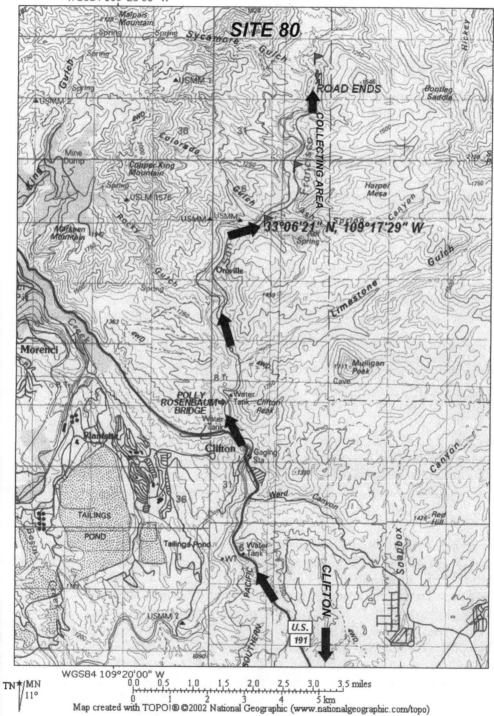

The San Francisco River Road was chiseled out of the steep sedimentary and igneous cliffs along the east bank of the San Francisco River north of Clifton. Since these cliffs are composed of soft, crumbly, and highly fractured strata, material is continually eroding onto the road to the consternation of the highway department and to the delight of mineral collectors. Landslides are common after heavy rainstorms. Agate, massive calcite, calcite crystal, selenite, obsidian, perlite, quartz crystal, chalcedony, geodes, and mollusk fossils have been found along this road. The calcitic material generally fluoresces red and the silicated rock fluoresces green. Although you can collect along the entire length of the road and beyond, the most fruitful areas begin at the 4.0 mile point. From this point on, lots of material has eroded from above and is piled behind barriers erected beside the road to catch it. Between the 4.5 and 4.8 mile points, you may find aragonite. Golf ball size chalcedony lined rhyolite geodes are eroding from a loose grainy stratum abut 50 feet above the road at the 4.8 mile point. Between the 5.6 and 5.8 mile points, large calcite crystal vugs and seams are visible in the cliff face. You can easily pick up crystalline calcite pieces that have fallen to the ground or you can try to extract pieces from the cliff wall. Resist the temptation to climb the cliff face to extract material from above. The cliffs are steep, unstable, and dangerous. Wait for it to fall. it is advisable to wear a hard hat while collecting under these unstable cliffs.

At the 6.1 mile point, you will come to a barrier marking the end of the road. You can park and hike farther up what used to be a road but is now more like a trail. Only a few hundred yards beyond the barrier, you will come upon an excellent example of the incredible geologic features that characterize this area. Slicing through the high perpendicular tuff cliff on your right is a fault about 30 feet wide that has filled with contorted bands of medium grey rhyolite. It looks like bands of twisted Christmas ribbon candy. Layers within this material fluoresce green. Several hundred yards farther on, you will come to a very high crumbly basaltic cliff. Rock slides cover the ground at the bottom of the cliff. Calcite geodes and vugs are visible in the soft rock. Since this rock is soft and crumbly, you can easily extract 1 – 3 inch specimens intact with your rock pick. Many are loose in the pulverized rock gravel piles at the bottom of the cliff.

The San Francisco River runs south from Whitetail Spring near Alpine, AZ through the forest to the Gila River south of Clifton. Coronado passed by it in 1540. The first known American citizen to visit the San Francisco River was a trapper named James O. Pattie and a companion, whose name is unknown, who trapped 250 beavers here in two weeks in 1825. He hid the pelts in what he thought was a safe place while he continued to explore upstream planning to retrieve them on the way back. Unfortunately for him, the local Indians discovered his cache and made off with his catch.

*G.P.S. coordinates taken 4.0 miles up the San Francisco River Road.

SITE 85

Fluorescent Calcite at the Summit Mine

Difficulty Scale: 4 – 4 – 3   Seasons: Fall, Winter, Spring

Global Positioning System Coordinates: 32° 52' 27" N, 118° 58' 22.6" W*

Geology: Middle Miocene-Oligocene Volcanic Flows and Tuffs

U.S. Geological Survey 7.5 Minute Topographical Map:
Goat Camp Spring (NM)

TO REACH THE SUMMIT MINE, follow the directions to Site 84, the Carlisle Mine, on page 218. See the Sites 84-85 map on page 219. But, when you reach the switchback do not follow the road around to the Carlisle Mine. Instead, continue on the road straight ahead of you that goes north. After going .6 mile, you will come to a fork in the road. Bear left uphill and precede another .7 mile and park

The ore shoot at the collecting area.

under the ore shoot. This is the collecting area. There is room to turn around farther up the road.

The tailings on the road and under the ore shoots contain pieces of massive red fluorescing calcite. Collecting is fairly easy. All you have to do is explore the area looking for specimens loose in the mine rubble and pick them up. You may dig through the rubble piles or bust open the larger rocks if you are so inclined. White, black, and colorless calcite is available here. Some quartz crystal vugs and seams as well as zones of banded agate are also present in limited amounts. The high mountainside above the ore shoot also contains collectable material. But, because of its steepness, it is a difficult area to prospect.

*G.P.S. coordinates taken at the ore shoot.

SITE 87

Fluorescents off U.S. Route 80

Difficulty Scale: 3 – 4 – 3   Seasons: Fall, Winter, Spring

Global Positioning System Coordinates: 32° 05' 21" N, 108° 85' 05" W*

Geology: Middle Miocene-Oligocene Igneous Volcanic Flows and Tuffs

U.S. Geological Survey 7.5 Minute Topographical Map: Steins (NM)

To reach the second New Mexico location along Interstate 10, continue eastward on I-10 about three miles from the exit to Site 86 (page 221) to exit 5 at Road Forks. Travel south from Road Forks on U.S. Route 80 10.5 miles. Turn left (south-east) on a narrow gated dirt road that leads .4 miles to a hill with a large tailings pile. You can see this destination from U.S. 80 as you approach the turnoff. You can park at the bottom of the hill right beside the collecting area.

The fluorescent material at this location is exceptional. Specimens containing red calcite, green willemite, blue-white scheelite, and white aragonite abound. There is a lot of collectable material here. Almost any rock you pick up will fluoresce at least one color and most will display two or three. Look for pieces that have a hard tan caliche crust on one side. Although this much maligned substance is found almost everywhere, it does fluoresce a very bright, pretty orange and adds one more color to the fluorescent array.

The turnoff from U.S. Route 80 to the collecting area

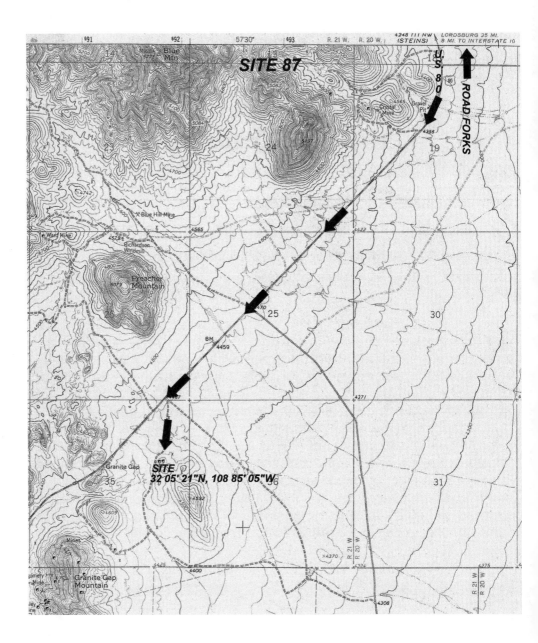

Collecting at this location is fairly easy. You can walk around the bottom of the large tailings pile and leisurely select rocks 2 – 3 inches across from the heap. Although generally earthy in appearance, many rocks here show a little semi-lustrous calcite, light brown aragonite seams, and subtle color variations suggesting the presence of other minerals present in the matrix. The hillside above the main pile is pockmarked with shallow prospect holes as well as open vertical shafts that expose more fluorescent material. Also present here is epidote and some dark blue botryoidal chrysocolla.

*G.P.S. Coordinate taken at the mine.

SITE 88

Fluorescent Minerals at the Hilltop Mine

Difficulty Scale: 7 – 6 – 4   Seasons: All
Global Positioning System Coordinates: 31° 59' 09" N, 109° 17' 25" W*
Geology: Jurassic-Cambrian Sedimentary Rock Metamorphosed to Quartzite
U.S. Geological Survey 7.5 Minute Topographical Map: Rustler Peak

The turnoff to the Kasper Tunnel showing the tailings pile on the mountain side.

FROM INTERSTATE 10, TAKE EXIT 340 and follow State Route 186 south from Wilcox 32 miles to S.R. 181. Turn left (east) on S.R. 181, go 3 miles toward the entrance to the Chiricahua National Monument, and turn right (south-east) onto Pinery Canyon Road (F.R. 42). Follow F.R. 42 5.6 miles and turn left (north-east) on F.R. 356, which is a narrow, bumpy, un-maintained dirt road leading up into the

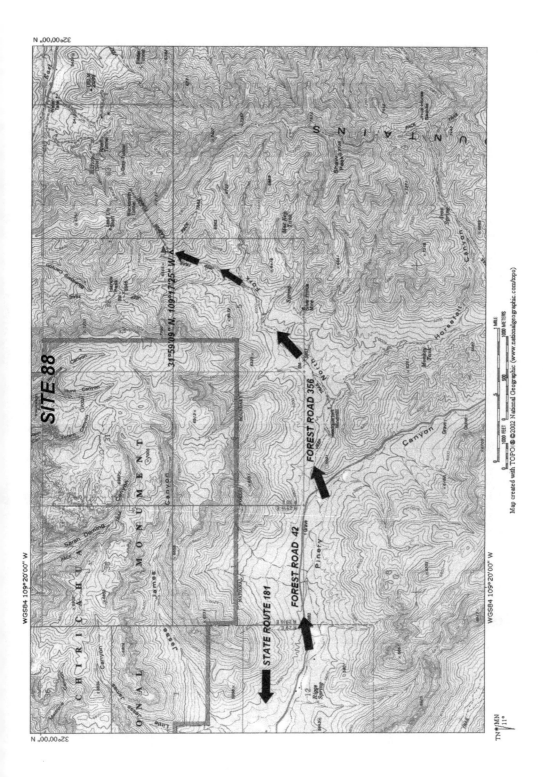

dense forest. High clearance is required. Follow F.R. 356 upward 1.7 miles to a turnoff to the right that leads up to the lower Hilltop Mine known as the Kasper Tunnel. The distance to the mine is .7 mile. At the .4 mile point, the road becomes steeper, rougher, and narrower. 4-wheel drive is advisable from here to the top of the tailings plateau at the mine where you can turn around and park. Or, you can choose to turn around and park at the .4 mile point and hike the rest of the way up. The Hilltop Mines farther up (the Rhem, Reider, Grey, Lead Lily, and Blacksmith tunnels) other prospects, and the site of the old town of Hilltop are at the end of a very rough 4-wheel drive road.

The first mining effort at what was to become Hilltop was done by Jack Dunn in the 1880's. In 1913, the Hilltop Metal Company acquired the property from Frank and John Hands who had purchased the property in the 1890's. With the rapid expansion of the mine by the Hilltop Metal Company, the town of Hilltop was officially established on the northwest side of Shaw peak in 1913. In 1917, a tunnel was completed through the mountain and the town site was moved to the northeast side of the mountain. Several houses, bunkhouses, a dance hall, pool hall, restaurant, and a garage for the town's only car, a six cylinder Cole, were erected. The post office opened in 1920. Productive mining lasted only until 1926. The mine changed hands several times in the 1930's and 40's with little success until operations ceased for good in 1949. The post office had already closed in 1945. The people drifted away and the town became a true ghost town.

When collecting here, use caution. The Kasper Tunnel tailings pile is steep and slippery. Trying to traverse it horizontally is difficult and dangerous. Although the safest way to approach it is from the bottom, it is also the most difficult because of the dense undergrowth. You can collect along the bottom of the pile and you can climb a short way up the face of the slope wherever you can gain a foothold if you are careful. Look for the dark colored rock among the predominantly light colored quartzite and limestone. Minerals reported here in the past include calcite, cerussite, sphalerite, scheelite, silver, willemite, wulfenite, and many others. The predominant fluorescent minerals to look for here are calcite (red), willemite (yellowish-white), and scheelite (bluish-white).

*G.P.S. coordinates taken on top of the tailings plateau.

Index

Actinolite 39
Adamite 73–74
Agate 7, 38, 68, 82, 115, 124, 128, 152, 166, 384, 386
Agglomerate 121
Ajax Mine 368–369
Ajoite 364
Allochemical 38
Alluvial deposit 192
Aluminum 39
Alunite 73–74
Amethyst 7, 80, 93–94, 125, 130, 180, 182, 220, 369–370
Ammonite 264, 318, 321
Amphorous minerals 29
Andalucite 39
Andesite 104, 118, 132
Anhydrite 327
Antimony 199
Apatite 327
Aragonite 7, 39, 112, 114, 199, 397, 381–382, 387
Aristides 255
Asbestos 169
Ash Peak 211
Augite 8, 60, 62
Aurichalcite 8
Azurite 8, 27, 321

Barite 9, 33, 59, 67, 68, 326, 347, 372
Barry M. Goldwater Range 144, 148
Basalt 62, 121, 333, 336

Basin and Range Province 2
BBC Mine 50, 53
Beryl 36
Big Bertha Extension Mine 76–77
Big Spar Mine 347, 352
Birefringent 28
Bivalve 270, 280, 282, 292
Black Pearl Mine (Maricopa Co.) 340
Black Pearl Road 156
Black Rock Mine 98
Black Silver Mine 344
Blastoid 266
Borate minerals 326
Boyer Gap 69–70
Brachiopodia 154, 183, 254, 263, 288, 292–293
Brochantite 62
Bryozoan 253, 263, 294

Cactus Queen Mine 141
Calcareous rock 39
Calcite 10, 32, 59, 64, 80, 106, 112, 114, 115–116, 120, 125, 130, 183–184, 189, 199, 323, 327, 333–334, 337–38, 340–341, 346–350, 352, 355–356, 358, 360–362, 370–374, 379, 381, 384–386, 387, 392
Calcium 254
Caliche 78, 328, 348
Cambrian 63, 67, 69, 162, 390
Cancrinite minerals 326
Carbon 254
Carboniferous 290, 320

Carnelian 130
Cast 254–255
Cenozoic 311
Cephalopoda 268
Cerussite 10, 183, 392
Chalcanthite 11, 220
Chalcedony 81–82, 115–116, 125, 132, 134, 148–149, 208, 337–338, 366, 338, 384
Chalcopyrite 11, 328
Chalk 108
Chasm Creek 284
Chlorite 39
Chromium 326
Chrysocolla 12, 51, 55–56, 59, 62, 64–65, 72, 90, 92, 94, 138, 142, 187, 220, 352,
Chrysotile 168–169, 177–178
Citrine 130
Clay 39
Cleavage 32
Cochise Mine 319
Coconino National Forest 296
Coelenterata 263
Color 27–28, 327–328
Colorado Plateau 2
Conglomerate 38, 333, 379
Contact metamorphism 39
Contact Mine 93
Copper Chief Mine 104
Copper Creek District 186
Coprolite 255
Coral 183, 253, 263, 284, 286, 296, 298, 299–300, 306
Coral Nuevo 165
Cottonwood Canyon 365
Cottonwood Gulch 121
Coyote Spring 208
Cretaceous 78, 87, 186, 198, 202, 304, 322, 379
Crinoid 183, 264–265, 274, 280–282, 288, 290, 292–294, 300–301, 303, 311, 315, 320–322
Critic Mine 86
Crook Tunnel 379, 381
Cruziana 255
Crystal habits 31–32
Crystal systems 30
Cummingtonite 12, 53–54
Cunningham Pass 87

Dactite 74
Dago Spring 308
Dana's *System of Mineralogy* 41–44
da Vinci, Leonardo 256
Dendrite 199, 221–222
Density 33–34
Descloisite 183
Desiccation 253
Devonian 321, 361
Diabase 321, 361
Diamond 32, 62
Diopside 39–40
Diorite 173, 186, 198
Dolomite 39, 284, 296, 298, 308
Druse 31
Dutchman Wash 47

Echnodermata 253, 264
Electromagnetic spectrum 325
Emerald 326
Empire Onyx Quarries 107
Epidote 13, 39–40, 62, 64–65, 69–70, 72, 92, 158, 387
Extrusive rock 35

Feldspar 39, 73, 76
Fire agate 210
Fluorapatite 327
Fluorescent 80, 82, 90, 120, 125, 142, 189, 196, 199, 216, 321, 323, 325, 333–392
Fluorescent activators 326
Fluorite 13, 27, 96–97, 135, 327

Fossil   3, 82, 253–332
Fossil classification   262
Fossil environments   265
Fossil formation   253
Fracture patterns   32–33

Gabbro   39, 62
Galena   14, 189, 328
Garnet   14, 39–40, 326
Gastropod   270, 280, 316, 319
Geode   78, 82, 134, 150, 165–166, 210–212, 297, 336, 338, 384
Geologic Column   256
Geological Time Scale   256
Geomystical creatures   256
Glauberite   11–112, 114
Gneiss   50, 53, 90, 101, 124
Gold   199, 328, 347
Gold Gulch   202
Gold Hill Mine   55, 58
Grandiorite   55, 58, 60, 205
Granite   55, 58, 60, 84, 87, 124, 128, 135, 138, 141, 156, 158, 173, 186, 198, 205, 378
Green Streak Mine   60
Grey Horse Mine   183
Griffin   255
Gypsum   48, 64, 111–112

Hematite   15, 50, 51, 54, 64–65, 68, 70, 72, 76–77, 87–88, 101, 130, 184, 328, 352, 390
Herodotus   255
Hexagonal   30
Hexavalent minerals   327
Hilltop Mine   390
Holocene   192
Homo sapiens   255
Hornblende   40
Hue   327
Hyalite   327
Hydrochloric acid   36

Hydrofluoric acid   36
Hydrothermal   36
Hydrozincite   366

Ichnology   254
Igneous rock   35–37, 54, 62, 76, 81, 115, 118, 128, 134, 138, 173, 186, 198, 336, 338, 382
Infrared   325
Intrusive rock   328
Iron   328
Iron pyrite   18, 158, 220, 254, 328
Isometric   30

Japanese Twin quartz crystals 188
Jarosite   64–65
Jasper   68, 101, 118, 120
Jurassic   63, 67, 69, 73, 76, 78, 81, 173, 195, 202, 304, 322, 379, 390

Keenan Camp   132, 134–135
Kyanite   132, 134–135

La Brea Tar Pits   253
Labradorite   28
Lava   96, 144–145, 148, 218, 336, 340, 355, 358, 382
Lead   199
Lead Pill Mine   135
Lepidolite   189, 51
Lime Hill   361–362
Limestone   107–108, 110–111, 153, 159– 160, 177, 183, 215, 222, 269–270, 272, 275, 278, 280, 282, 284, 288, 290, 296, 298, 301, 304, 308–310, 313, 316, 319, 322, 361–362, 382, 392
Limestone Canyon   272
Linné, Carl von   256, 262
Lithium   347
Little Giant Mine   87
Luster   27, 29

Maggie Mine 36, 336
Magma 35, 39
Magnetite 328
Malachite 15, 58–59, 62, 64–65, 72, 84, 86, 88, 92, 94, 138, 140, 142, 159–160, 189–190, 220, 321, 352
Malfic rocks 39–40
Mammoth Mine 84
Mammoth Spar Mine 347
Manganese 98, 199, 327, 347
Marble 16, 69–70, 108, 162, 164, 195–196
Marine fossils 278, 313
McCracken, Jackson 129
McCracken Mine 128
Mesozoic 311
Metamorphic rock 39–40, 50, 53–54, 58, 60, 63, 67, 69, 90, 92, 101, 162, 180, 195, 347, 365–366
Metamorphic rock protolith 178, 390
Mica 17, 121, 122
Mimetite 16
Mineral environments 35
Mineral Material Sales Permit 6
Mineral Mountain 370
Minerals of Arizona 1
Mingus Mountain 101, 281
Miocene 50, 53, 55, 58, 60, 96, 104, 107, 111, 118, 121, 132, 144, 148, 165, 183, 205, 208, 211, 215, 218, 275, 333, 336, 340, 355, 358, 382, 385, 387
Mohs Scale 32–33
Mold 114, 254
Mollusca 183, 254, 264, 269, 287, 297, 311, 315, 384
Monochromatic minerals 27–28
Monoclinic 30
Mudstone 39, 111, 160, 183, 215, 275, 288
Mule Mountains 316
Multicolor minerals 27–28

Munsell Book of Color 328

Nail Canyon 269
Nanometer 325
Nautiloid 264
New England Mine 138, 140
Newsboy Mine 347, 349

Obsidian 29, 80, 144, 150, 166, 214, 384
Oklahoma Mine 372
Oligocene 50, 53, 55, 58, 60, 115, 144, 148, 165, 205, 208, 355, 358, 382, 385, 387
Onyx 2, 39, 199, 173–174
Opal 29, 152, 166, 215
Orthochemical 38
Orthorombic 30
Optical properties 27
Oxidation zone 36

Paleontology 255
Paleozoic 159, 278, 281, 284, 296, 308, 310, 382
Papagoite 364
Pegmatite 31, 87, 152, 158
Pelitic rocks 39,
Peloncillo Mountains 210, 221–222
Peridotite 39
Permian 269, 272, 275, 287, 293, 313, 316, 319
Petrified Forest National Park 254
Petrified wood 210, 253
Phenocryst 206
Philips Mine 168
Phoenix and Yuma Mines 63
Phosphorous 254
Physiographic provinces 2–4
Plant fossils 275–276
Pluton 35
Porifera 263, 293
Porphyry 73, 76, 186, 198, 205

Precambrian   261
Prince Mine   356, 358
Proterozoic   84, 90, 93, 101, 124, 128, 135 138, 141, 156, 158, 168, 176, 180, 347, 365, 368, 375
Protoceratops   255
Protolith   39–40
Pseudomorphs   114, 254
Pyroclastic rocks   148, 165
Pyroxene   40

Quartz   2, 18, 29, 32, 51, 62, 64, 68, 70, 73, 76, 77, 80, 88, 92, 94, 101, 106, 122, 125, 130, 134, 140, 152, 158, 166, 173, 182, 188, 202, 210, 220, 369–374, 384
Queen of Sheba Mine   350

Ramsey Mine   78
Rawhide Mine   141–142, 384
Rhyolite   78, 80, 104, 118, 132, 150, 152, 165–166, 177, 212, 321
Rockhound etiquette   5–6
Rosiwal Scale   32–33
Ruby   326
Rugosa   263, 300, 304
Rutile   29
Ryan Ranch Road   162

San Francisco River Road   382–384
Sandstone   39, 159–160, 183, 215, 275, 284, 296, 298, 301, 379, 382
Sardonyx   210
Scapolite   326
Scheelite   19, 33, 125, 326, 366, 375–376, 387, 392
Schist   50, 53, 90, 101, 347, 365, 368, 375
Schorl   36, 122
Scilla, Agostino   256
Scott Mine   340
Scythian   255

Sedimentary rock   36, 38, 47, 63, 67, 73, 78, 96, 107, 115, 153, 159, 160, 168, 176, 183, 202, 259, 261, 269, 281, 290, 308, 344, 382
Selenite   19, 47, 74, 96–97, 130, 135–136
Septarian nodules   38
Serpentine   20, 39, 169
Shale   308, 382
Shattuckite   364
Siderite   21, 104, 106
Signal City   124–125
Silicon   130, 254, 309
Silver   199–200, 328, 347, 392
Site Difficulty Scale   24
Sodalite   326
Sphalerite   392
Spinel   326
Steno, Father Nicolas   256
Stenofiber   255
Strata   256
Stratigraphy   256
Streak   28
Sugarloaf Peak   73
Systema Naturae   262

Talc   32, 39
Taphonomy   253
Tellurium   199
Tenacity   34
Terrigenous   38
Tertiary   198
Tetragonal   38
Thunder eggs   38
Tombstone Hills   198
Tonto National Forest   296
Topaz   326
Trace fossil   308
Travertine   21, 107–108, 199
Triassic   47
Triclinic   30

Trigonal 30
Trilobite 253, 266, 280
Tourmaline 36, 189
Tuff 81–82, 96, 121, 144, 208, 218,
    221 333, 340, 355, 382, 385, 387
Tungsten 326
Tungsten King Mine 375
Twin Buttes 355

Ulexite 28
Ultraviolet 325
United States Mine 159
Uraninite 327
Uranium 326–327
Uranyl 327

Vanadinite 21, 183–184
Verde Antique Marble 39, 69
Volcanic Bomb 38

Wallastonite 39
W A Ranch Well 205
White Peak Mine 356, 360
Willemite 22, 196, 366, 387, 392
Willy Rose Mine 356, 360
Woodpecker Mine 180, 182
Wulfenite 121, 183, 392

Xenophanes 255

Zinc 199, 327
Zircon 327

ABOUT THE AUTHOR

NEIL BEARCE is the consummate prospector of earth's treasures. From colonial artifacts in New England, to deep sea diving for seashells in the South Pacific, to gold mining in California's Sierra Nevada, to the rocks and minerals of Arizona, he has found the beauty and mystery of nature's creations irresistible.

While earning a Bachelor's degree at the University of Maryland, he studied under Dr. Gordon W. Prange, author of *Tora, Tora, Tora,* and had the privilege of contributing to that renowned work. He went on to earn a Master's degree at the University of Akron and a second Master's degree at Golden Gate University, where he graduated Summa Cum Laude.

His interest in rocks and minerals began in his youth while excavating the ruins of a revolutionary war era foundry near his boyhood home in Massachusetts. Among the ancient cannon balls and musket shot were pieces of quartz and iron ore. The bug had bitten. The next thirty years found him in Micronesia, Southeast Asia, Europe, the Mediterranean, and the Near East where he collected black and orange coral, jade, star sapphire, and meerschaum. He also toured the Tiger Balm jade collection in Singapore, the Idar Oberstein gem mines in Germany, and the quarry in Carrara, Italy, where Michelangelo procured the marble for his statue of David.

Since 1984, he has resided in Tempe, Arizona. He is a member of the Leaverite Rockhound Club where he served as Trip Chairman. He is also a member of the Arizona Mineral and Mining Museum Foundation and the Tucson Gem and Mineral Society. His first book, *Minerals of Arizona: A Field Guide for Collectors,* was published in 1999.

Keep on truckin!